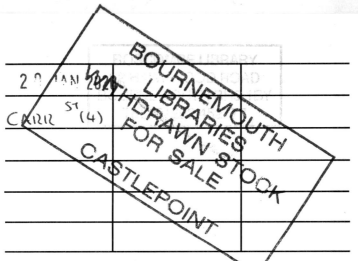
- You can return this item to any Bournemouth library but not all libraries are open every day.

- Items must be returned on or before the due date. Please note that you will be charged for items returned late.

- Items may be renewed unless requested by another customer.

- Renewals can be made in any library, by telephone, email or online via the website. Your membership card number and PIN will be required.

- Please look after this item – you may be charged for any damage.

JEFFREY MASSON

The Secret World
of Farm Animals

VINTAGE

3 5 7 9 10 8 6 4 2

Vintage
20 Vauxhall Bridge Road,
London SW1V 2SA

Vintage is part of the Penguin Random House group
of companies whose addresses can be found
at global.penguinrandomhouse.com

 Penguin
Random House
UK

First published in Vintage in 2005 with the title
The Pig Who Sang to the Moon

penguin.co.uk/vintage

A CIP catalogue record for this book is available
from the British Library

ISBN 9781529111026

Printed and bound in Great Britain by Clays Ltd, Elcograf S.p.A.

Penguin Random House is committed to a sustainable future
for our business, our readers and our planet. This book is
made from Forest Stewardship Council® certified paper.

For Leila, Ilan, and Manu

Contents

Preface

Three years ago, my family and I were visiting Auckland, New Zealand, when we heard about a pig who lived on a beach just fifteen minutes from downtown. This pig was famous; school children came to visit, she had been proposed for mayor, and her neighbors were fiercely divided between those who thought a pig living on the beach was a bit of magic and others who feared she would devour their children. We found the beach, but Piglet, as she was called, had moved to a macadamia-nut orchard farther north and was no longer on view.

To make a long story short, we met her guardians, the wonderful Tony Watkins and his equally wonderful partner Helen, and wound up buying a house on that very beach. We heard many stories about this amazing pig who liked to go for a swim early in the morning when the sea was at its calmest, and who enjoyed having children sit on her side, as long as they gave her a tummy rub before leaving. She was immaculate, well-mannered, sensitive, intelligent, and kind to strangers. When we finally met her, we could see that you could not ask for a better neighbor or ambassador for farm animals. Her emotional life was particularly near the surface. She always let you know what she was feeling; most of the time it was obvious from the smile on her face, especially when she was swimming or playing with her small human friends.

But there were more mysterious aspects to her as well. She was sensitive to music and liked to hear the violin played. She especially

seemed to enjoy music on the beach at night when there was a full moon. Tony took a picture, quite recently, of her making the sweetest sounds during a night of the full moon, as if she were actually singing to the moon. The picture at left, of Piglet singing, is photographic evidence of her special affinity for music, water, night, and moon. It is another reason to believe that many animals—pigs foremost among them—may have access to feelings humans have not yet known. Perhaps if we listen carefully enough to the songs that Piglet and her cousins sing at night to the moon, we may yet learn about emotions that could bring us a new and utterly undreamed of delight.

In my books about animals, I have usually devoted a few pages to speculate about emotions that animals have which humans may lack. Of course it is purely speculative, but there are moments in the lives of everyone who lives in great intimacy with another species, when we are suddenly aware that the animal is slipping away from us, entering a realm to which we do not have access. She gets a distant look in her eyes, or her face lights up with a kind of joy we seem not to know. At such moments, I am convinced that if I could just learn to listen better, or were just a bit more open to the unknown, I might slip into that same space and learn about emotions of which I at present know nothing.

There are many people who have undergone great suffering who seem to possess knowledge of the deepest recesses of human emotion unavailable to the rest of us. They may want to impart it, but we often cannot hear. Strangely enough, farm animals strike me in the same way. To some people, of course, it sounds absurd to compare the lives of farm animals to survivors of human tragedies. The more, though, that I learn about their lives in factory farms, the more I am convinced that the analogy is not as far-fetched as it might appear. Farm animals—in spite of or perhaps because of the fate that invariably awaits them—seem able to retain their capacity for deep feeling, including, miraculously, their love for us. It is my hope that this book will alert you to the existence of another kind of animal life, one that has been with us for thousands of years, but which we rarely truly see. I want Piglet to sing to us as she sings to the moon—perhaps a song of mourning, of loss, of sadness, but perhaps also a song of joy, of compassion, of love.

Acknowledgments

My editor in England, Tony Colwell, died while he was editing this book. We had been friends for many years, and he was a wonderful editor. I cannot say how sorry I am he is no longer here to enjoy the results of his many hours of work on this book that was close to his heart. "At last," he told me when I said I wanted to write a book about the emotions of farm animals, "the balance will be redressed a little. It is high time."

As usual, Nancy Miller, editor-in-chief at Ballantine, has done more than any author has the right to expect, going over draft after draft after draft, always improving the text and clarifying my thinking. I would not dream of writing any book without Nancy's help.

Elaine Markson, my agent for twenty years, has been her usual friendly, helpful self. All writers deserve such an agent.

Kim Sturla, the cofounder of Animal Place, a sanctuary near Davis, California, held my hand throughout the writing of this book. I visited her wonderful sanctuary many times and benefited from the lucidity of her mind and the compassion of her heart. She introduced me to Gene and Lorri Bauston, who founded Farm Sanctuary, in upstate New York, one of the largest farm animal sanctuaries in the world. There's another Farm Sanctuary in Northern California. These people's example, and their activism has inspired countless people to take up work on behalf of these wonderful and all too often neglected animals. I am in awe of all they do.

Jim Mason put me up for a week while we visited farms in the Midwest. He knows probably more than anybody else about farm animals and how they suffer, and feels for them deeply. I learned a great deal from being with him, listening to him, and working with him. He is a wonderful friend and a superb author. I read and reread his books.

Matthew Scully, a better writer than I, went through the whole manuscript and improved it greatly.

I visited Karen Davis, the founder of United Poultry Concerns, and was able to see firsthand her love for chickens and other fowl. Nobody knows more about them, or cares more deeply for them, than Karen. I highly recommend her two wonderful books about chickens and turkeys. They altered my thinking.

Joyce D'Silva, the director of Compassion in World Farming in England, took time out of a busy schedule and drove me to sanctuaries and farms throughout England. She was a delightful companion, gentle, sensitive, and knowledgeable. What her organization achieves in Europe has become an inspiration for many. Her colleague, Jacky Turner, who did research and taught physical chemistry at the University of London, did some brilliant research for me and was a wonderful person with whom to discuss farm animal issues.

I thank the people of New Zealand for the friendly welcome they accorded my family and me, and for their faith in a more just world. I am waiting for the day when everyone realizes that animals must be part of that dream.

Without the warmth, depth, and love of my wife, Leila, I could never have written on this topic. I dedicate the book to her and our two children, Ilan, who loves all animals and refuses to ever eat them, and Manu, who at eighteen months has already learned to be gentle to all living creatures. It is my fervent hope that they represent the beginning of a new consciousness and conscience toward everyone with whom we share this fragile planet.

I do not have the space to thank every person whom I met on the many animal sanctuaries I visited, but I want them to know that without their example, their hard work, and their dedication to the animals in their care, I would not even have begun to think about writing a book like this.

Introduction:

A Good Life

Walk into any bookstore and say that you want to read something about farm animals and you will be sent to the children's section. There is nothing for adults about farm animal behavior, let alone farm animal emotions. That is why I set out to write this book. It turns out that the farm animals I studied (chickens, pigs, cows, sheep, and goats) have, as might well have been expected, complex emotions, among them love, loyalty, friendship, sadness, grief, and sorrow. The domesticated animals who live on our farms are very little removed from their wild ancestors and therefore have all the emotions that belong to those wild animals who live under conditions of freedom. This means that confinement is going to be all the more painful for farm animals, conflicting as it does with emotions that evolved under far different conditions. (We can only imagine the emotions of farm animals subjected to factory farming, where it is impossible to perform *any* meaningful natural behavior.) There is then no difficulty answering the question of whether these animals are happy. An animal, much like a human, can only be happy when it is living under conditions that allow it to express its natural behavior and feel the emotions that accompany that behavior. Considering the wild ancestors of farm animals allows us to answer many questions that were deemed pointless or unanswerable in the past. These answers may make some people uncomfortable, but they can give us, at last, the kind of insight into the personalities and needs of farm animals that has been missing for centuries.

If a farm animal has a good life and that life ends in a painless death and the animal is used to feed people, is that wrong? Many people would answer that it is not. But I think it is worth asking first just how anybody knows what a "good life" is for a farm animal. Of course, we all have some idea what this might mean. However, few people outside of industry apologists would be prepared to argue that the average dairy cow leads a happy life. Consider the cow whose calves are removed at birth, and who is then milked intensively for a few years. She is kept almost permanently pregnant to keep her milk flowing—while not allowed to keep her calf. Finally, old before her time, her usefulness for giving milk in decline, that same cow is killed, long before her natural life span has been reached. Can we say that this cow has led a happy life?

The position I take in this book is a radical one. Still, I feel it is so simple and logical I do not know why it has not been taken before, and by large numbers of people. To the extent that you prevent an animal from living the way he or she evolved to live, you are creating unhappiness for that animal. All farm animals, it turns out, from chickens to cows, have evolved to have offspring and guide them through their infancy. An animal who has given birth and is then deprived of the opportunity to give expression to her inborn maternal urge cannot be happy. Think about it: If you were made pregnant against your will, and then your child was taken from you and served the next night for dinner, would you be happy? If you think a cow never gives a second thought to her missing calf, ask any farmer how long a newborn calf and her mother call for each other. One farmer told me that as long as they can see each other, they will call until they are hoarse, indefinitely.

I think it is wrong to raise animals for food. I just do not believe that anybody will take care to give an animal a "good life" if the point of that life is to end up as a meal on the table. It is too easy to cheat, too tempting *not* to search out what makes for a good life for any particular animal. Adequate, tolerable, bearable: these are the adjectives we are happy to use for the conditions under which farm animals live. They are clearly not adjectives we would aspire to for our own living conditions. Imagine if you were to ask a boarding school director what kind of life he contemplated for your enrolled child, and he replied: "It will be adequate." You might easily surmise that your child's life would not be good.

Even those who honestly attempt to give a farm animal a good life know that this animal's life will never be optimal. Feelings the animal is capable of are not given scope in any kind of life that gives short shrift to his or her evolutionary past. So yes of course, it is better to allow a chicken to live the kind of outdoor life with others of her kind that she would live had she the choice. But surely we can all understand that it is even better to let the chicken live her allotted time, rather than serving her up for dinner.

Where, though, can farm animals live the kinds of lives they were meant to live? Farm animal sanctuaries are the only places to my knowledge where animals can do so. A farm sanctuary provides a safe home for animals who have been rescued from slaughterhouses or from abusive situations. There they are able to live out their lives without ever being exploited for their flesh, their hide, their eggs, or any other body product. They are cared for, they are respected, and they are loved by volunteers who are honored to be with them. Chickens have evolved over millions of years to live a certain kind of life. The kinds of changes that occur naturally over hundreds of thousands of years have not yet taken place during the short period that chickens have been domesticated. Chickens did not evolve for the benefit of humans. Domestication was never their purpose. We breed chickens, of course, for our benefit, but it will become apparent to anybody who takes the trouble to look that we have not succeeded in breeding out of chickens that which makes them chickens and not dinner.

I've noticed that when I tell people over a meal that I am writing about the emotional lives of farm animals, they give me an odd smile, as if I had said something slightly ridiculous. Then they dive into their steak or their lamb or their chicken or their pork, seemingly not the least bit curious about the lives of the animals they are consuming for dinner.

Not *what* are you eating, but *whom* are you eating is the question on my lips. Should the topic of suffering on such a vast scale be considered ridiculous? Is it ridiculous to care that 24 million chickens are killed *every twenty-four hours* in the United States? That is almost 9 billion (nine thousand million!) chickens for the year 2002. As for pigs, there are 268,493 killed every twenty-four hours in the U.S. alone, for a total of 99,236,800 in the year 2002. We will never know the true number of animals slaughtered for food in the world per year, but the number is almost beyond imagining. We know, for example, that 40 billion chickens (6 for every person on the planet) are killed for food every year in the world, but that does not include chickens in countries that provide no figures nor does it include backyard slaughter or unauthorized slaughter of any kind. So the real number is unknown, but immense.

Why is it generally considered ridiculous to point out that each and every one of those animals killed had a mother, almost all had siblings, and surely some were mourned by a parent or missed by a friend? Even though they were bred to be killed, their emotional capacities were not altered by such breeding. They had memories, they suffered, and they grieved. There is little justification for making a comparative scale of suffering where "human" is weighted and animal is given little weight. To be concerned about one kind of suffering does not mean that you must have no interest in another, or that you think one is somehow more significant or more terrible than another. Jean Améry, a survivor of the Holocaust, did not want to be forced to find similes to describe his suffering: "Pain was what it was. There's nothing further to say about it. Qualities of feeling are as incomparable as they are indescribable."[1] And so the suffering of almost all farm animals is unique, particular, mostly beyond language to describe or explain. If we give it no thought, and yet eat them for our meals, are we not morally blind, ethically dumb, and humanly remiss?

Around ten thousand years ago, as the great Ice Age came to an end, human beings gave up the life of hunter-gatherers and began farming, but this was not a simple matter of choice. Climate change had brought with it a scarcity of water, and thus of game. It made sense to settle down near an oasis or other water source and plant crops of wheat and barley. Before that time, Desmond Morris tells us in *The Naked Ape*, we had "killed and eaten almost any animal you care to mention." From a study of prehistoric remains we know that about half a million years ago, at one site alone, "humans were hunting and eating species of bison, horse, rhino, deer, bear, sheep, mammoth, camel, ostrich, antelope, buffalo, boar and hyena." Even before crop planting got under way, we found ourselves up against one or two serious competitors for edible prey—particularly the pack-hunting wolflike ancestor of our domestic dog. With a sharper sense of smell and hearing, and being more skilled at herding and driving prey at high speed, the hunting dogs would often rob less adept humans of an essential part of their daily diet. Exactly how a symbiotic bond was formed between dogs and humans is unknown, but Dr. Morris theorizes that puppies were originally brought home as food, but that it soon became apparent that they made good watchdogs at night. Moreover, dogs would accompany men on their hunting trips and even regard them as their own leaders, and cooperate fully in the hunt, since they now saw themselves as members of the human pack.

The horse has had almost as crucial an effect on the development of farming as the dog. The horse was the last of the five most common livestock animals to be domesticated, and as Juliet Clutton-Brock, an expert on the history of domestication, points out, "as a species it has been the least affected by human manipulation and artificial selection."[2] Originally a source of food for prehistoric man, it did not take people long to realize that the horse evolved to cover great distances in a short period, and this speed and strength could be exploited as a means of human transportation and farming the land. Doing so transformed the way human beings lived.

Fossil evidence from Jericho (the oldest walled city in the world) shows sheep and goats were domesticated there around nine thousand years ago. Goat remains have also been found in a cave in Palestine known as El-Khiam, going back to ten thousand years ago. The telltale physical signs of domestication—reduction in size, shortened faces, and compacted teeth—are evident in the remains found at these sites.

Many people have a distorted view of what farm animal domestication refers to. Let me clarify: Every farm animal has been domesticated. By farm animal, I mean those animals who live on a farm. That is a bit of a misnomer now, because most farm animals, at least in the United States and in Europe, do not live on farms. They live in artificial sheds, barns, or stables that resemble human prisons. We concentrate the animals into these places because it is convenient and profitable to do so. It has nothing to do with, and has never had anything to do with, the comfort, safety, or health of the animals.

Some people have objected to the term "farm animal" on the grounds that these animals are not there by choice. We farm them, and so it would be more accurate to call them farmed animals—the emphasis placed on the doer, us, rather than the done to, them. Were the phrase not so clumsy, I would have preferred to use "farmed animal" through the book. When I say "farm animal" I mean an animal who has been farmed. It is, alas, an us/them situation. The origins of the sense of "an other" may very well owe its life to the idea that early humans had when confronted with the ultimate "other": an animal. Animals were primarily seen by them as either prey or predator. Their desire to turn these "others" into animals less alien marked the beginnings of domestication. The idea was to make the animals more docile (smaller, with less strength), slower, so that we could catch them more easily, and more willing to accept servitude.

In the case of dogs and cats, you could make the argument that domestication was mutually beneficial. They got companionship, security, food, warmth, playmates, and even love. We got many of the same things back (minus food—since rarely did humans eat the animals with whom they played and shared their homes). But in the case of farm animals, the relationship never approached anything like an even exchange. We took their eggs, their milk, their flesh, their skin, their work, and in exchange they got, as far as I can see, the short end of the stick. While we protected them from predators, this was only because we, a more powerful predator, had already earmarked them for our own exploitation, invariably ending with their personal extinction. How it can be argued that this is to protect them has always struck me as odd.

One of the mysteries about the earliest contacts between domestic animals and humans is whether it was the animals' usefulness or our attraction to them as companions that was the original impetus for domestication. It is not clear when humans first began to milk other species, either the goat or the cow, but milk was probably not the main reason they were domesticated, since so many people in the world outside of Europe and North America were at the time (and are today) lactose intolerant. A small kid goat just a few days old, able to eat and wander about on her own, is an easy animal to catch. Once caught, she would behave much like the young of any animal—helpless, harmless, and cute in the extreme. It is conceivable that people ten thousand years ago as much as today found themselves won over by the cute features of the animal—big eyes, floppy ears, playfulness—which evolved so that the baby would endear herself to parents and other members of her species as a way of ensuring her survival. It's not surprising that it works with us as well. The humans of ten thousand years ago were identical to us in genetic heritage, intellectual capacity, and brain development. No doubt they saw the similarity between domestic animals and wild animals, and between both kinds of animals and themselves. This worldview was given the stamp of science by Charles Darwin a little more than a hundred years ago, although we seem, somewhere along the line, to have lost the connection or to have grown so arrogant, or so human-centered, that we can no longer see it.[3]

George Orwell's classic novel, *Animal Farm*, is generally considered to be a political fable about totalitarianism and Russia. Literary critics and ordinary readers alike have seen his tale of farm animals as merely a device, an engine for the story. Orwell, however, saw it in another light, explaining in a preface written for the Ukrainian translation, that the story came to him when he saw a little boy, perhaps ten years old, abusing a carthorse. He was struck with the force of a revelation "that men exploit animals in much the same way as the rich exploit the proletariat." He went on to explain that he turned Marx's fundamental insight on its head: "I proceed to analyze Marx's theory from the animals' point of view. To them it was clear that the concept of a class struggle between humans was pure illusion, since whenever it was necessary to exploit animals, all humans united against them: *the true struggle is between animals and humans.*"[4] Although I have added the emphasis, there is every reason to suggest that Orwell truly believed this. In fact, at the beginning of the book, Orwell has Major tell the animals on the farm, "No animal in England knows the meaning of happiness or leisure after he is a year old. No animal in England is free. The life of an animal is misery and slavery: that is the plain truth." Is this not a deliberate echo of the most famous statement made by Jean-Jacques Rousseau—that man was born free, and everywhere he is in chains?

The philosopher Stephen Clark once remarked that human slavery only began with the arrival of farming. Hunter-gatherers took no slaves, nor did they very often engage in war, since, without property, there was little to fight about. It was farming, and domestication, with their innate sense of things that belong to certain people and not others, that saw the growing importance of status and the beginning of war.[5]

The comparison between slavery and animal domestication is not new. Like so many of our Western attitudes, it goes back to Aristotle, who wrote in *Politics*:

> For all tame animals there is an advantage in being under human control, as this secures their survival. . . . By analogy, the same must necessarily apply to mankind as a whole. . . . these people are slaves by nature, and it is better for them to be subject to this kind of control, as it is better for the other creatures I have mentioned.

War and hunting, Aristotle went on to claim, are part and parcel of this philosophy and both must be waged against animals and "those men whose nature it is to be governed."[6] How convenient these laws seem to be for the men of Aristotle's class. Such attitudes die a slow death.

There is a common view that farm animals would not even exist unless we bred them: better for them to lead a confined life than no life at all. It is commonly asserted that animals like cows, pigs, sheep, goats, chickens, ducks, and geese benefit merely by being allowed to exist. The British philosopher and fox-hunting enthusiast Roger Scruton, for example, makes the curious argument that "Young animals have been slaughtered without compunction from the beginning of history," as if slavery and racism and the abuse of women were not also practiced from the beginning of history. Since when does the practice of something over time confer moral rectitude? He goes on to say, "Most of the animals which graze in our fields are there because we eat them." They could still be there, grazing in our fields at a sanctuary, if we did not eat them, but of course their numbers would be fewer. But it does not seem a sound philosophical point to say that somebody or something owes its very existence to our desire to exploit, as if that confers some special moral right upon us. He concludes, "It seems to me, therefore, that it is not just permissible, but positively right, to eat these animals whose comforts depend upon our doing so."[7] But their comfort *need* not depend upon our eating them; we could decide that their comfort was important in and of itself, without reference to any benefit that we might derive. *This* is the true moral position, not the one derived exclusively from self-interest.

Some writers take perverse delight in listing faculties humans possess and animals lack. Marc Hauser, professor of neuroscience at Harvard University, in his new book *Wild Minds*, attempts to make the point that animals do not possess moral choice in the way that humans do. He ends his argument with a passage that, in spite of myself, I found profound. He says that if animals are moral agents, they must understand the norms of action and emotion in their society and "have the capacity to engage in a revolution when their rights have been violated." He suggests that no subordinate animal has ever thought about changing the system, overthrowing "the normative responses and feelings that define life in a primate group." So while a subordinate animal may feel that he is being treated unfairly, Hauser says he knows "of no instance where an animal has attempted an overthrow of societal norms. No subordinate has ever built up a coalition of support to derail the system."[8] This is an original observation, and one well worth thinking about. However, we must put it in context: what percentage of humans has ever raised revolutionary ideas? Is not truly revolutionary thinking rare in the history of ideas? But more important, why compare the abilities of humans and animals? Bias invariably creeps in, for there would not be much point to the exercise except as a means of upholding human superiority. Fewer and fewer scientists care to engage in this futile pursuit.

When we think about farm animals, it is important to keep in mind that the purpose of their existence is almost entirely defined by their death or exploitation. They exist to die or to be used. We breed them to kill them or profit from them—not to provide them with a way of fulfilling their destiny for a happy life. No amount of philosophical blather can get us past this immovable rock of, may we call it human treachery? We facilitate the birth of farm animals because and *generally only because* we intend to kill them sooner or later, and generally sooner (lambs, chickens, calves). If they can be exploited in some other way first (dairy cows, wool-bearing sheep, milk-giving goats, egg-laying hens) their lives may be prolonged, but generally not beyond the point where they cease to be of economic benefit. Although there are farmers who care about their animals, I suggest that the majority do not think like James Granger, the vicar of Shiplake in Oxford, who in 1772 said that we should show kindness to farm animals, "our poor servants . . . and especially when they are advanced in years, and worn out with drudgery."[9]

We acknowledge that certain animals see better than we do, hear more, smell more acutely, are stronger, faster, more agile, and so on. This does not make them better or superior, just different and worth studying and valuing. Why not when it comes to the emotions as well? I am sure that the ability to feel terror may well be as advanced in many other prey species as it is in humans. The need for friendship may be more powerful in many other species than it is in ours. The love of children may be highly developed, even more than in our own, in many species that form pair-bonds. (The British philosopher Mary Midgley, in a conversation with me, thought this to be likely.) Fidelity in geese, patience in cows, playfulness in lambs, the list could continue for pages. Just imagine how rewarding it could be to contemplate the emotional superiority of an animal when it comes to compassion or love: to think that we have something to learn.

ONE

Pigs
Is
Equal

An old English adage claims, "dogs look up at you, cats looks down on you, but pigs is equal." There is some truth in the folk wisdom of this saying, which has been ascribed to different people, including Winston Churchill, but nobody is sure who said it first. Pigs are more or less the same size as human beings and resemble us in many ways. The organs of pigs are so similar to our own that surgeons have resorted to pig heart valves for replacing a patient's aortic or mitral valve.

There is a wonderful quote from W. H. Hudson, the great naturalist who lived for some time in Argentina, that perfectly describes the pig's attitude toward us:

I have a friendly feeling towards pigs generally, and consider them among the most intelligent of beasts. I also like his disposition and attitude towards other creatures, especially man. He is not suspicious or shrinkingly submissive, like horses, cattle and sheep; nor an impudent devil-may-care like the goat; nor hostile like the goose, nor condescending like the cat; nor a flattering parasite like the dog. He views us from a totally different, a sort of democratic standpoint, as fellow citizens and brothers, and takes it for granted that we understand his language, and without servility or insolence he has a natural, pleasant camaraderie, or hail-fellow-well-met air with us.[10]

The fact that pigs will become extremely friendly with humans, given half a chance, is something of a miracle, considering how we have almost invariably treated them. Perhaps pigs themselves are aware of our resemblance and so regard us more as cousins than members of a completely different species. Unlike dogs, pigs don't seem to have a critical period after which they can no longer be socialized. Handled with affection, even an adult pig might well become as friendly as a dog who has always lived with the family since puppydom. This shows remarkable trust and flexibility on the part of the pig. The one big difference between pigs and dogs is the way we treat them. We play with our dogs, take them for walks, and romp with them. We rarely do the same with pigs.

One has to wonder why the pig came to be despised by both Jews and Muslims. Was it merely the flesh of the pig that was distrusted, or the pig itself, as an animal? By and large people have believed the former, claiming that because pig meat was so easily prone to spoiling and trichinosis, the consequent human diseases led them to avoid the meat and thereby censor the animal. But the late F. E. Zeuner, a leading expert on domestication, rejects this view, pointing out that pork is no more likely to spoil than any other meat in a hot country, and in any event there are tropical islands where pork is the main meat eaten. He proposes instead an interpretation having to do with the people who raised pigs. Unlike cattle, pigs cannot be driven, and therefore the pig is only valuable to the settled farmer. The no-mad, who always felt superior to the farmer, "came to despise the pig as well as the farmer who bred it." The religious prohibitions seem to have been transferred from the people on to the animal, one they "themselves could neither breed nor keep."[11] But then why would this not apply equally to chickens? Could it be that they are smaller and more transportable?

In most parts of the world today, we cannot own another person in the way that we can own an automobile. The law is also increasingly taking the view that a human cannot "own" an animal companion either.

This became evident when a wealthy man in Philadelphia sought to have his two dogs euthanized after his death. In a surprising victory for the views of the movement for the rights of animals, a United States federal court decided that animal companions cannot be owned, and therefore could not be disposed of at will as if they were merely chattels. The logic then (as now) is that living beings can never be property. As early as the nineteenth century, Henry Bigelow, professor of medicine at Harvard University, was writing: "There will come a time when the world will look back to modern vivisection in the name of Science, as they do now to burning at the stake in the name of Religion."[12]

It is undeniable that we humans share a great deal in common with pigs, though people have been reluctant to acknowledge the similarities. Like us, pigs dream and can see colors. Also like us, and like dogs and wolves, pigs are sociable. (On warm summer nights pigs snuggle up close to one another and for some unexplained reason like to sleep nose to nose.) The females form stable families led by a matriarch with her children and female relatives. Piglets are particularly fond of play, just as human children are, and chase one another, play-fight, play-love, tumble down hills, and generally engage in a wide variety of enjoyable activities. As Karl Schwenke points out in his classic 1985 book, *In a Pig's Eye*, "Pigs are gregarious animals. Like children, they thrive on affection, enjoy toys, have a short attention span, and are easily bored."[13] He reports that when pigs were put into a small pen, as they are on most farms, "their world was instantly narrowed to each other, the food, and the sty, and as they grew, their world became smaller and smaller. The tedium of their existence soon became apparent: they were lethargic, exhibited ragged ears, had droopy tails, and rapidly acquired that dull-eyed glaze that swineherds associate with six- or seven-year-old breeding hogs." Much like children, piglets do not develop in a normal way when they are deprived of the opportunity to engage in play.

Kim Sturla, of the northern California animal sanctuary Animal Place, tells me that pigs express friendships with other pigs a variety of ways: vocalizing, body language, who they sleep with, explore with, hang out with during the day. Some pigs, Sturla says, are friendly with certain pigs because they arrived at the sanctuary about the same time. Juveniles will play with each other, and immense patience is demonstrated with new piglets. One can witness the interaction and affection when pigs greet each other, snout to snout, sometimes with love grunts—soft, wispy, open-mouthed greetings given when a pig is in heat, feeling amorous, or maybe just feeling sweetly affectionate. Pigs can also be cliquish: an older new arrival may not easily find acceptance.

Like humans, pigs are omnivores. Though they are often fed garbage and eat it, their choices—if allowed—would not be dissimilar to our own. Sturla tells me that when she offers her pigs mango or a head of broccoli, they will always take the mango. She explains that they have a sweet tooth and a pastry will always win over a healthy vegetable. Remind you of somebody? They get easily bored with the same food. They love melons, bananas, and apples, but if they have had them for a few days, they will set them aside and eat whatever other food is new first. We don't often think of pigs and cleanliness in the same breath, but pigs, if permitted, will be more fastidious in eating and in general behavior than dogs. When offered anything unusual to eat, a pig will sniff at it and nibble gently. About 90 percent of their diet in the wild is plant-based, consisting of fruit, seeds, roots, and tubers. In fact, a study of what fruits pigs routinely eat, conducted on one of the Indonesian islands, found that they would eat more than fifty varieties. Perhaps this is why of all animals their flesh most resembles human flesh, which is somewhat disconcerting when you consider that more than 40 percent of all meat raised in the world is pork.

Like people, pigs avoid extreme temperatures. Since they have sweat glands only on their noses, it is important that they do not overheat. Water is not effective in cooling them down because it evaporates quickly, whereas mud provides evaporative cooling over a much longer period of time. This is why pigs, like elephants, need to roll in mud. Mud protects their sensitive skin from sunburn, dangerous to a pig, and also from flies and other parasites. It is not, then, that pigs are dirty; quite the contrary. Never will a pig defecate near its sleeping or eating quarters. Fastidiousness is one of a pig's most salient characteristics. Kim Sturla has repeatedly seen old arthritic sows waking up early in the morning, getting their stiff bodies up with enormous effort, then dragging themselves through deep mud to walk a long distance away from the barn before they would urinate. We can only imagine the suffering involved when pigs are confined in such a small space that they refuse to foul their own stalls, as *Compassion in World Farming* has documented and others have noted.

And if we find them sometimes difficult, because they can, like humans, have tantrums, it simply means that they are prone to powerful emotions. Early behavioralist Ivan Pavlov, after a month of fruitless attempts to obtain gastric juice from a vociferous pig, declared: "It has long been my firm belief that the pig is the most nervous of animals. All pigs are hysterical."[14] But let us bear in mind that this same comment has been made numerous times about women and, in both cases, it is an example of pure ignorance. What is clear is that pigs are much like us in ways that matter. There is nothing shameful in recognizing the similarity.

The resemblance extends to the expression in the eyes of pigs. Many people have found it disconcerting to look into the eye of a pig. This is because one gains the startling impression of seeing another person looking back at you. Pigs have small, rather weak eyes and appear to be squinting, as if they are trying to get a better take on the world. They seem often to wear a wistful look. Dick King-Smith, the writer who created *Babe* (turned into the much-loved film) and who used to be a pig farmer,[15] said on a television show, "Many times I've looked into a pig's eye and convinced myself that inside that brain is a sentient being, who is looking back at me observing him wondering what he's thinking about." When I recently visited Carole Webb's Farm Animal Rescue in Cambridge, England, I was introduced to Wiggy, a gigantic boar (a male pig) weighing nearly a thousand pounds. As I came into his stall, he was busy picking out soft hay with which to line the straw in his self-made bed. He grunted when I walked in, looked up, and fixed me with his eye. It was uncanny, like meeting a person in the street whom you feel you know but cannot place. I looked away for a moment, embarrassed by the naked intimacy of his glance.

Juliet Gellatley, in her book *The Silent Ark*, describes visiting a factory-farm shed where she saw a large male boar, "his huge head hanging low towards the barren floor. As I came level with him he raised his head and dragged himself slowly towards me on lame legs. With deliberation he looked straight at me, staring directly into my eyes. It seemed to me that I saw in those sad, intelligent, penetrating eyes a plea, a question to which I had no answer: 'Why are you doing this to me?'"[16] If we are to consider pigs as sentient beings with intelligence and a full range of emotions, perhaps we should feel guilty when a pig gives us that look knowing he will soon be off to his death.

Not many people have been able to get inside one of the factory hog farms that blight the Midwestern and Southern United States today. Matthew Scully did, though, and he has written one of the most scathing yet compassionate books about animals in the entire literature: *Dominion*. I cannot recommend it highly enough. He watches an expectant mother "nose at straw that isn't there to make a nest she'll never have for another litter she'll never raise." And he reminds us that "in exchange for their service they get exactly nothing, no days of nurturing, no warm winds, no sights and sounds and smells of life, but only privation and dejection and dread." After visiting one such place in America, Scully writes: "How does a man rest at night knowing that in this strawless dungeon of pens are all of these living creatures under his care, never leaving except to die, hardly able to turn or lie down, horror-stricken by every opening of the door, biting and fighting and going mad?"[17] He reminds us that "a child playing with a toy barnyard set, putting all the little horsies and piggies outside the barn to graze, displays a firmer grasp of nature and reality than do the agricultural experts . . ."

The Tamworth is one of the oldest breeds of domesticated pig. The Tamworth Two captured the imagination of Britain in 1998 when these two pigs escaped from a truck taking them to slaughter, burrowed under a fence, swam the Ingleburn River, and fled into a thicket from which they could not be induced to come out. Seen as a worthy escape, it won massive public sympathy. Even the slaughterman, Jeremy Newman, who first sighted them five days later, admitted: "You can't be sentimental in this business, but I say good luck to them. I reckon they got more sense than we have, they showed a lot of initiative when they escaped. As soon as they caught sight of me, they made off as fast as their legs could carry them."[18] For some people, it was the first time they realized that a pig does not want to die. There were hundreds of offers to provide the pigs with a safe haven for the rest of their lives. They now live in an animal sanctuary where they need never again fear the slaughterhouse. Given the urgency of their escape, it seems likely that the pigs sensed what lay ahead for them. How they did so is another question.

Some people are uncomfortable with the notion of collective guilt, that we should bear for acts committed by another. But feeling to some extent complicit is not necessarily a pointless emotion; it can move us to moral responsibility and to activism. Vegetarians often feel less compromised in their relations with animals they do not eat.

I do not know whether pigs know that they will be killed, but I do know that the screams of a pig being killed bears an uncanny resemblance to human screams and that people who have heard them are unnerved. One wife of a farmer I knew in the South became so despondent that she told her husband she would no longer participate in the killing and would leave him unless he found a way of farming without killing animals. They are now peach farmers.

Much like humans, every single pig is an individual. (I can never say this too many times about every farm animal simply because we humans have a tendency to forget this important fact—no doubt because we think individuality is a human prerogative.) Some pigs are independent and tough, and don't let the bad times get to them. Others are ultra-sensitive and succumb to sadness and even depression much more readily. The example that comes to my mind is a story Kim Sturla told me about Floyd. This was a pig living in pig heaven at the wonderful Northern California Farm Sanctuary, along with all his siblings. For complicated reasons, he needed to be sent to Animal Place when he was nine months old. Nothing wrong with this: it doesn't get much better than Animal Place, nor could there be a more loving mom than Sturla. She knew right away that she had a sensitive soul on her hands. Floyd went straight into the barn and would not emerge. He would not eat, though Sturla gave him the choicest grapes, which she hand-fed him, one grape at a time. Sturla introduced him to Penelope, a young, sweet, and submissive pig. He would not play. He just whined, as if weeping from sorrow. He went into what looked like a deep depression. It seemed this pig would not survive; he was giving up. Sturla could not figure it out. Finally, Diane Miller, who had been Floyd's caregiver while he was at Farm

Sanctuary, came up to see what the problem was. The moment he saw her, his whole demeanor changed. He smelled her with what looked like relief on his face, suddenly alive with emotion, and squealed with delight. This same pig who would hardly move at all, raced to the Econoline van and jumped into the back, ready to head for home. That was all he wanted: to go home to the pigs he knew and loved. As soon as he was back at Farm Sanctuary, not a trace of his depression remained. His heart was sick away from his own home. Sturla could not have been more loving toward him, but Floyd just missed his family. Surely we can recognize the similarity to sensitive humans and sympathize with this equally sensitive pig.

Many people who are around pigs a good deal remark upon how gentle pigs can be when they are well treated. Much like dogs, pigs seem to have a great capacity for gratitude and to know when they are liked. Evidently they are capable of reciprocation.

Pigs know their names, and, again like dogs, they wag their tails when happy. They are also capable of rescuing their owners. On Thursday, March 9, 2000, the *London Daily Telegraph* reported the remarkable rescue of Dee Jones by her pet pig, Pru. Dee and her husband, Simon, kept Pru as a pet because they thought her brighter than other pigs: "I watched her as she moved bales of hay so she could stand on them and climb up to open the gate with her snout," said Jones. How glad she was that she did. When walking with her sheepdog in Whitland, West Wales, Jones fell waist-deep into a bog. She panicked at first, but when Pru came over she put the dog leash around the pig and told her: "Go home, go home!" Pru obeyed, pulling Jones out of the mud and saving her life.

Being human, we like to believe we can understand such unusual happenings. What is the explanation? Since a human in distress makes a sound very similar to a pig in distress, Jones's pig clearly understood that the situation was dangerous and was intelligent enough to obey the order to return home. Here the pig's curiosity, her ability to use insight, and her natural affection for people to whom she feels close, came together to save the life of someone dear to her.

It is hard to accept instinct as the only valid explanation for the extraordinary behavior of Lulu, a 200-pound Vietnamese potbellied pig living at Animal Place. These animals are direct descendants of the Old World wild pig that had spread across Eurasia for over forty thousand years. Their behavior is basically identical to that of a barnyard domestic pig and if we think of them as gentle and kindly this is because we behave that way toward them. This is what the following story demonstrates: Joanne Altsmann was in her kitchen one afternoon, feeling unwell, when Lulu charged out of a doggie door made for a 20-pound dog, scraping her sides raw to the point of drawing blood. Running into the street, Lulu proceeded to draw attention by lying down in the middle of the road until a car stopped. Then she led the driver to her owner's house, where Altsmann had suffered a heart attack. Altsmann was rushed to the hospital, and the ASPCA awarded Lulu a gold medal for her heroism. Altsmann knows in her bones that Lulu's sixth sense saved her life.[19]

What did it require for the pig to do what she did for Altsmann? Obviously a commitment to her friend, some awareness of how to bring help, the desire to do so, and the ability. It seems unlikely that all this could have happened without conscious awareness, yet we are unwilling to credit the pig with a thought like: "Oh dear, Joanne is in serious trouble. At whatever cost to my own well-being, I must bring her the kind of help that can save her life." Of course, to do so would also be in the pig's best interest. But if one pig can do this, then we must assume that other pigs can as well. If a pig is able to feel this profoundly and make such self-sacrificing decisions, what does it say about the casual way we slaughter them? What must their emotions be at such a time? Every account I have seen of butchering stresses how much the pig is aware of what is happening and how much of a fight it puts up to escape. A butcher told me: "You can see it in their eyes, they know what is about to happen to them."

Maybe in the case of humans whom they consider part of their family, pigs will respond to distress as they would with another pig. Pigs are gregarious and will come to the rescue of another pig who is being harmed. A typical example of the loyalty of pigs to other pigs in distress comes from the Uganda Game Department. A male African giant forest pig had been "wounded by a spear and was partially paralyzed in the hind legs. He was limping badly and four of his family were close around him, trying to help him out of a mud hole and making a great noise about it."[20] The special cry of a pig in distress is an immediate signal to all other pigs in the area to rush to its aid. They do it so consistently and with such clear beneficent intentions that it strikes everyone who sees it as perplexing. Despite this, scientists who study wild pigs are reluctant to characterize this with anything that smacks of human values. With human beings, we would call this compassion; in pigs, we prefer to say it is purely instinctive.

Some people like the idea of keeping pigs as pets but are understandably put off by the size of the barnyard pig. Hence the fascination with the tiny and extremely rare wild pygmy hog, which barely survives on the borders of Bhutan and Assam in India. So small is this miniature pig that you could hold several in the palm of your hand, much like the newly discovered tiny primate *Eosimias*, or "dawn monkey," which lived forty-five million years ago in a humid rain forest in what is now China and was no bigger, when fully grown, than a human thumb. Why there is such a desire to have an animal that you can hold in your hand is not clear, but evidently Vietnamese potbellied pigs like Lulu became so popular as pets because they were believed to be miniature. Potbellied pigs were introduced in the United States in 1985 and promoted as the perfect house pet. People were told that a pig was easier to housetrain than a dog (true), and that these pigs would stay small and adorable and didn't require a lot of room (not true). These pigs do not stay small.[21] Since the average adult potbellied pig weighs 150 pounds, it was not long before growing pigs were abandoned, much like dogs purchased on a whim at Christmastime only to become a nuisance when the holidays are over. Maybe it is only an accident of history that most people consider dogs as pets and pigs as dinner, that the dogs sleep at their feet, and pigs outdoors. Were it not for the size of the pig, perhaps it could have been the other way around.

In general, the more we know about something, the more we care. It would seem that the closer a person comes to a pig, the greater the value he or she begins to place upon pigs in general. Familiarity breeds respect. If farmers have failed to find this value in pigs, it may be because they refuse to see it since they know they must kill them.

Can a pig forgive? The Norwegian author Bergljot Borresen writes about a mountain farmer from western Norway who learned an important lesson from his big sow. Extremely fond of people, she was in the habit of lying with her head on top of the metal railing around her pen when there were people in the barn. Those passing her would speak kindly to her and pat her head. One day the farmer had to repair a rotten board on the floor of her stall. She was curious and continually nudged him as he worked. Annoyed, he smacked her with his hammer. "I should not have done that, for immediately she took my thigh into her big mouth and locked it completely between her jaws, though she did not bite. She probably only wanted to warn me not to do such a thing to her ever again. She found it intolerable that I would do something unkind to her."[22] This story demonstrates a remarkable series of abilities on the part of the pig. She had a sense of justice and of the consequences of breaking certain rules of behavior, but also of making allowance for someone who could not be expected to have mastered all the fine points of porcine etiquette. Then there is forgiveness and compassion—not traits we generally think of as prominent among pigs.

That some humans have loved pigs even when, for hundreds of years, society has told them they ought not, is abundantly clear. There is a passage in George Eliot's *Scenes of Clerical Life* (written in 1857) that deserves to be better known:

> Such was Dame Fripp, whom Mr. Gilfil, riding leisurely in top-boots and spurs from doing duty at Knebley one warm Sunday afternoon, observed sitting in the dry ditch near her cottage, and by her side a large pig, who, with that ease and confidence belonging to perfect friendship, was lying with his head in her lap, and making no effort to play the agreeable beyond an occasional grunt.
>
> "Why, Mrs. Fripp," said the Vicar, "I didn't know you had such a fine pig. You'll have some rare flitches [bacon] at Christmas!"

"Eh, God forbid! My son gev him me two 'ear ago, an' 'he's been company to me iver sin.' I couldn't find i' my heart to part wi'm, if I niver knowed the taste o' bacon-fat again."

"Why, he'll eat his head off, and yours too. How can you go on keeping a pig, and making nothing by him?"

"Oh, he picks a bit hisself wi' rootin', and I dooant mind doing wi'out to gi him summat. A bit o' coompany's meat an' drink too, an' he follers me about, and grunts when I spake to 'm, just like a Christian."[23]

One of the first books devoted entirely to swine, William Youatt's *The Pig*, published in 1847, took a similar perspective: "In a native state swine seem by no means destitute of natural affections; they are gregarious, assemble together in defense of each other, herd together for warmth, and appear to have feelings in common ... How often among the peasantry, where the pig is, in a manner of speaking, one of the family, may this animal be seen following his master from place to place, and grunting his recognition of his protectors."

On the other hand, society's attitude toward pigs has become embedded in some ugly phrases. "Your place looks like a pigsty." "Don't be such a road hog." "He is a filthy swine." Although older expressions, such as "You can't make a silk purse out of a sow's ear," have a negative ring to them, the symbolic pig, it seems, only came to predominate culturally in proportion as fewer and fewer people saw real living pigs.[24]

The actual pig of nature was banished from human consciousness as vast farms on which the pig had only one purpose—meat—

became ever more prevalent. Even on traditional family farms, pigs were rarely seen as companions or even animals with a dignity of their own (unless these farms were products of fantasy, like the great literary classic *Charlotte's Web*). There were bound to be some exceptions, since a farmer who came to know his pigs as individuals was less likely to disregard completely their welfare. As people learn more about the remarkable abilities and sensibilities of pigs, attitudes are beginning to change. We have to credit the emotions of a pig in a movie, *Babe*, with a surge of people turning away from eating pork and a dawning recognition that pigs have been unjustly maligned in our language and in our literature.

Like many of us, Charles Darwin had nostalgic views about English farms. His biographers, Adrian Desmond and James Moore, note that he was always pleased to see "an English farm house & its well dressed fields, placed there as if by an enchanter's wand." These farmhouses were surrounded by well-kept gardens of flowers, fruit, and vegetables; threshing barns; forges; and "in the middle was that happy mixture of pigs & poultry which may be seen so comfortably lying together in every English farm yard."[25]

Yet Darwin refers to a sow that produces less than eight piglets at birth, that she "is worth little, and the sooner she is fattened for the butcher the better."[26] Actually, here Darwin is quoting someone else, yet this is one of the few occasions where Darwin does not give a source for a quotation—presumably because this was such a generally shared opinion that it was considered common wisdom. A pig is for slaughter and therefore the only purpose of a sow was to give birth to as many piglets as possible for the express purpose of their slaughter. The sow, of course, has no such purpose in mind. She will expend enormous effort in providing a well-lined nest for her young, and take great pains to see to it that they are hidden from danger. A pig values the lives of its young much as we value ours. So what really is the purpose of a pig?

In 1962, Dr. K. C. Sellers, director of the British Animal Health Trust's farm livestock research center, was reported by *Farmer's Weekly* as having ". . . pointed out that pigs were kept to make money, *as carcasses*, and one should not get over-sentimental about them." Is it better to give them a more interesting life if the end is the same— slaughter? He thought not. In the same journal, a farmer reported, "In the last war, I hired a derelict house and farm building and put about 100 pigs therein. Part of one wall in the house had collapsed but the staircase was intact and upstairs there was the bedroom to which the pigs had access. The pigman reported that there seemed to be competition for the bedroom every night and that in the daylight hours they would chase each other up and down the stairs. *I never had pigs do better than that lot.* [Italics in original; he means they were healthy and happy.] I have come to the conclusion that our stocks need variety of surroundings and that gadgets of different make, shape and size should be provided and that, like human beings, they dislike monotony and boredom."[27] In almost all zoos today there is great concern for enrichment, for enhancing the everyday lives of the animals there with anything that will simulate their natural environment. Too often this is nothing more than a pathetic old tire thrown into the bears' cage, but the fact that zoo management thinks of such things is proof that we have moved beyond our earlier notions of animals as devoid of emotional needs.

If a pig exists only to provide man with meat or leather or some other by-product of its body then we need not concern ourselves with whether a pig is happy or fulfilled or living the right sort of life. But if we admit that pigs have emotional lives, if we are willing to look upon them as more than meat, that they have other reasons to live, then we can ask important questions from the point of view of the pig: What makes a pig happy? How is a pig fulfilled? When is a pig leading a good life?

Yes, we have bred pigs into existence, and we continue to breed them, but the idea that we therefore *own* them is not unlike the curious argument used by Southerners in the past to justify slavery. Domestication protects animals, this argument goes, just as some Southerners once argued that slavery protected blacks. Otherwise, as Judge Joseph Henry Lumpkin of the State Supreme Court of Alabama put it, "These children of the sun would perish if brought into close contact and competition with the hard and industrious population . . . northwest of the Ohio."[28] Or it would be like parents today claiming that their children's rights are irrelevant because the children would not even exist had their parents not brought them into the world. Among humans it was long ago abandoned as a moral principle that because you brought something into existence, you were master of its fate.

While in the United States, pigs continue to be kept in horrendous circumstances—such as much too small crates—more and more countries in the European Union have laws about how pigs and other farm animals may or may not be treated. These laws exist only because of the belief that pigs and other farm animals are capable of experiencing suffering, and that it is incumbent upon anyone who has anything to do with these animals to lessen that suffering to whatever extent is possible. In the U.K., sow stalls were often cages so narrow sows could not turn around or might not even be able to stand or lie down. There were also tethers, open crates in which the sows were tied by a neck collar and chain. These practices are no longer permissible in the U.K.

Of course human compassion and understanding still have a long way to go. When the Codes of Practice were first issued in 1969, Sir Julian Huxley and nine fellow scientists wrote to the *London Times*: "It is obvious that behavioral distress to animals has been completely ignored. Yet it is the frustration of activities natural to the animal which may well be the worst form of cruelty."[29]

Questioning the purpose of a pig can be as complex as the same questions about human existence. Freud went so far as to claim that anyone who asked what was the purpose of life by the very question gave evidence of mental illness! Yet every one of us constantly asks this question in one guise or another. What am I doing with my life? What do I *really* want, what do I care most about, what is it that I wish to achieve? It is doubtful that these questions bother animals, so following Freud's reasoning they are not as neurotic as we are, which seems to be something he, like so many others, believed. (Witness Walt Whitman's famous line about wanting to live with animals because "they do not lie awake in the dark and weep for their sins.") Nonetheless, we can ask these questions on behalf of animals, and even answer them to some extent. We might say that the purpose of the life of a pig is simply to be a pig, and a pig is happiest when a pig is doing what a pig evolved to do. We learn what a pig evolved to do by looking at what wild pigs do in natural conditions.

The domestic pig, the wild pig, and the eastern Asiatic wild pig are all in the same family, Suidae, which includes the bearded pig, warthog, pygmy hog, and babirusa. By human standards, many of these animals are considered ugly, which may be where we derive some of our anti-pig prejudices: The warthog, an African wild pig, has large tusks, as does the male babirusa from the Indonesian island of Sulawesi. Like all wild pigs, they are good swimmers and are almost entirely nocturnal. The extraordinary feature of the babirusa is the double set of "tusks." The upper canine teeth grow upwards, break through the skin of the upper jaw, and curve up and around toward the eyes. As yet, scientists have been unable to discover their purpose: The tusks are not strong enough for fighting, and they do not appear to be attractive in courtship. Once again we see how correct Montaigne was to maintain that we do not admire what we cannot understand.

Pigs and hippopotamuses belong to a mammalian order, the even-toed ungulates. The Eurasian wild pig, or boar as it is called, is the antecedent of the overwhelming majority of domesticated and feral pig populations. It has one of the widest geographic distributions of all land-based mammals and is found on all continents except Antarctica and on many oceanic islands. Due to hunting, it is now extinct in the British Isles, Scandinavia, and most of the former U.S.S.R. The pig was domesticated many times, in many different places. A number of authors have noted that the pig is one of those animals psychologically preadapted to domestication, predisposed as they are to cooperate with humans. It is far from clear, however, that people always domesticated pigs to use them as a source of meat.

In ancient Egypt, pigs were used for threshing. Bundled grain was spread in a contained area and trampled on by the sharp hooves of pigs. They were considered extremely valuable. In an illustration on the granite sarcophagus of Taho (about 600 B.C.) in the Louvre Museum, we see a boat with a domesticated pig on board guarded by two monkeys standing upright and holding what look like whips. Evidently they are protecting their precious charge. In some cultures, pigs were used for planting grain, since the holes their hooves created in soft earth were the ideal depth for seeds. Indeed, there were (and are) pigs who can do many of the tasks normally assigned to dogs. Thus domestication expert F. E. Zeuner reports an 1807 story of Slut, a hunting pig:

> Of this most extraordinary animal will be here stated a short history, to the veracity of which there are hundreds of living witnesses. Slut was bred in the New Forest and trained by Mr. Richard Toomer and Mr. Edward Toomer, to find, point and retrieve, Game [sic] as well as the best pointer; her nose was superior to any pointer they ever possessed, and no two men in England had better. Slut has stood partridges, black-game, pheasants, snipe and rabbits in the same day, but was never known to point a hare. When called to go out shooting, she would come home off the forest at full stretch, and be as elevated as a dog upon being shown the gun.[30]

Intelligence does not imply worthiness; in other words, it should not matter, from an ethical perspective, how intelligent a particular species or even any particular individual is—after all, we don't shoot a human being who is not doing as well as his contemporaries at school. Nonetheless people tend to comment on the particular intelligence of these "horizontal humans," as they are sometimes called because they are so much like us. In 1898, a professor of physiology in Montreal, Wesley Mills, compared the pig favorably with the dog:

> What would the dog be today if he had, for hundreds of years, been valued only for his flesh, and kept exclusively to

be fattened for food? The hog is charged with being dirty, stupid, and obstinate. Why should an animal, overburdened with flesh and fat, and consequently a sufferer from the heat of summer, be so much blamed for betaking himself to a pool even if muddy? Man is largely responsible for enforcing conditions involving filth on the hog.[31]

Eurasian wild pigs, the ancestors of the domestic pig, will run away in a fanlike formation when a predator chases them, then drop back and encircle the foolhardy animal who gave chase to them in the first place. It is a dangerous situation to be surrounded by a herd of angry, powerful wild boars. When hunted by humans, feral or wild pigs revert to a nocturnal existence, apparently as a deliberate adaptation. They seem to understand that they will be easier prey during the day, but harder to hunt at night. Undoubtedly, they are resourceful, clever animals. How different their lives are now, and how far from its natural self the pig has come.

In many American factory farms, pigs are routinely sedated and kept in dark or semidark giant sheds so that all they can do is eat and sleep for twenty-three out of twenty-four hours. All the *joie de vivre* is driven out of the pig. Fattened to immobility, tails cut, teeth removed, their natural instinct to investigate frustrated by the cold concrete floor, their desire for order destroyed by being forced to exist in a small pen, their sense of cleanliness ruined by being forced to urinate and defecate in their sleeping space, something no pig would ever do in nature—they have been utterly metamorphosed. So heavy are they that many find it difficult to stand on their own legs. Piglets are taken from their mothers when they are about three weeks old and put into "nursery" pens—where, of course, no nursing takes place—with metal bars and concrete floors. They move to "growers" and "finishers," and at about six months, when they attain the "slaughter weight" of 250 pounds, they are killed. They are subject to any number of diseases because of the overcrowding—arthritis because they cannot exercise, salmonella infection (present in one-third to one-half of all farms), epidemic gastroenteritis, porcine parvovirus (PPV, the most common infectious cause of reproductive failure). The mothers fare no better. They are confined in small pens or metal gestation crates (two feet wide) after pregnancy and four months later are transferred to farrowing crates where they barely have room to stand and lie down. They are denied straw bedding (too expensive). Unable to exercise or even move, they become very heavy (the point, of course) and subject to crippling leg disorders. Psychologically, they become "neurotic," as the breeders call them, biting the bars of the crates, in a sitting position much like dogs, but looking dazed, showing all the signs of mourning the loss of their babies. Then it is time for the slaughterhouse. Being highly strung animals, many of them suffer from porcine stress syndrome, also known as malignant hyperthermia, a deadly increase in body temperature leading to death. An industry expert states, "These are the animals that start shaking and die of a heart attack."[32]

No longer curious, gregarious, investigative, or self-reliant, the factory farm pig has had all its natural attributes removed from it. It is almost as if quite on purpose every single thing the pig is meant to do has been subverted, suppressed, even extinguished. The pig's life has been distorted, perverted, deformed, contorted beyond recognition. They are not allowed to live any part of their lives in the natural world outside. They never see the sun.

We take it for granted that humans deserve freedom. Traditionally this has meant freedom from oppression, but now increasingly it has come to mean that we will have the freedom to become whatever we want to be. This is a new kind of freedom, and one could argue that humans want to be what they were designed to be, to realize their full potential.[33] Animals who evolved to live a good part of their lives outdoors, investigating, wandering, and engaging the natural world, cannot lead a natural life if they are denied this capacity. To wish to give these same opportunities to animals seems to many humans absurd—or worse, perverse. But with humans, too, we should bear in mind that 150 years ago, when the world looked very different, with empires dotting the globe, human subjugation was the rule, not the exception. The "masters" would have mocked any insistence that others deserved the same liberties that they enjoyed. Today, we look back at that period aghast.

In earliest times, farmers would simply let their pigs go to the forest at night to eat. This was the cheapest way to fatten them. It wasn't done out of kindness, but the result was that these pigs led a much more natural life than most pigs today. Are the people who keep pigs in such awful circumstances today less compassionate than the earlier farmers? I would say that farmers today keep themselves in ignorance of the needs and true nature of pigs precisely because to know would put their conscience in a terrible bind. Willful ignorance of this kind is no better than complicity, though if the modern pig farmer rarely thinks about the pig as an animal with a nature, with needs of its own, it may be due less to a lack of compassion than lack of thought.

In *Jude the Obscure*, Thomas Hardy noted the accusatory gaze of a pig that was horribly slain:

> "The dying animal's cry assumed its third and final tone, the shriek of agony; his glazing eyes riveting themselves on Arabella with the eloquently keen reproach of a creature recognizing at last the treachery of those who had seemed his only friends."[34]

Almost every farm animal, when given the opportunity, reverts back to its forebear and takes on the characteristics of its wild progenitor. The pig is no exception. Given half a chance, it will live under feral conditions as a wild boar. Darwin, in his chapter on "reversion" from *The Variation of Animals and Plants Under Domestication*, writes that pigs "have run wild in the West Indies, South America, and the Falkland Islands, and have everywhere acquired the dark color, the thick bristles, and great tusks of the wild boar; and the young have reacquired longitudinal stripes."[35] Earlier in the text he had explained, "the tusks and bristle reappear with feral boars, which are no longer protected from the weather." This demonstrates very clearly how close to wild pigs domesticated pigs—kept like bacon-producing machines—really are.

Richard Guy of the Real Meat Company in Wiltshire, England, explained to me that his meat is what he calls "kind meat." He tries, he says, to give his animals the best possible life before he kills them. The notion of a kind death (that is, by the way, what euthanasia means), for an animal, is strange and puzzling. Clearly, these deaths are in no way mandated, they are not inevitable; in other words, they are not necessary. That much is clear. All you need do to make them unnecessary is to say once and mean it: These deaths are not necessary. I do not *have* to eat meat.

But this is still a minority position. What about people who choose to eat meat, but still feel uncomfortable knowing that these animals are leading lives of misery? Matthew Scully, in his recent book *Dominion*, says that Robert F. Kennedy Jr.—son of the slain senator—took the principled position that it is fair to eat meat if the animals have lived a good life. Scully thought that this was an acceptable position, or at least preferable to utter indifference.

Richard Guy and I argued about this in a good-natured way, neither of us changing our views or succeeding in getting the other to do so. I had to admit that "his" cows (I use quotation marks because the idea of owning a cow still grates on me—I just don't see how anyone can own a living being, human or animal) seemed well cared for. They were free to roam, they were not fed antibiotics, and they were leading better lives than any intensively farmed animals would ever live.

I was also impressed with the total transparency policy of Guy's company. They had nothing to hide, and anybody was welcome to visit the slaughterhouse they used or any of the farms where their animals lived. I took Guy up on this offer and asked to visit a pig farm.

I went with Joyce D'Silva, the head of Compassion in World Farming, a European organization that has successfully campaigned for reforms in the way farm animals are treated. The farm we visited was run by a man who was not cruel to the pigs under his care. He gave them fresh straw daily. They were not kept in horrendous conditions. But they were confined indoors in small and crowded holding pens. My first and lasting impression was: This is no life. The response of the farmer, when I told him my views, was well, of course not, how could it be? He was concerned with the welfare of these pigs only in the technical sense of that term. They had to be healthy so that he could make a profit. The pigs were there not for their own benefit, but for his. They were, he told me, working for him. They were employees. (It was not possible to get the pigs' view of this, but I am pretty certain they would have major disagreements; needless to say, they had no choice in the matter.) I asked the farmer whether he liked pigs. The question took him aback. "Never thought of it that way," he told me. "They are just a means to an end for me." The end was money. "I want my pigs to be healthy," he told me.

"What about happy?" I asked.

He hesitated.

Clearly this farmer had not thought much about what makes a pig happy. I did not think his pigs could possibly be happy. They were living in crowded conditions, they did not get to be outdoors, in fact, and they could not even see the outdoors. They could not form friendships, wander, protect their young, experience moonlight, build nests; in short, they were not living the lives pigs were meant by evolution to live. How, then, could they possibly be happy? If you were to provide a maximum of happy conditions, acres of hills and valley and natural ponds to wallow in, well, then the next thing you know you wouldn't end this little idyll with a trip to the slaughterhouse, would you? Once you make the transition to think about *their* happiness, not your own, then it becomes clear that being killed for food is not conducive to porcine happiness.

I know there are people who ask what would happen to pigs if we did not raise them for slaughter, apparently worried that they might become extinct. It seems to me that there is something a little bit hypocritical on the part of people who benefit from the death of pigs to worry about what would happen to them if we no longer wanted them for meat. They would no doubt revert to a wild status and would eventually manage their own numbers in accord with their environment, exactly as every other wild animal does to the extent that humans allow them to. The lives that pigs lead on a factory farm is not anyone's idea of a happy life, neither that of a pig nor that of any human being who has ever witnessed such a life or given it any thought.

People have argued that we have to keep pigs in these conditions because if there is no profit in raising them, if we do not eat them, *and* make a profit raising them, they will cease to exist. Roger Scruton, the British philosopher and foxhunting apologist, goes so far as to claim, "I find myself driven by my love of animals to favour eating them. Most of the animals which graze in our fields are there because we eat them . . . it seems to me, therefore, that it is not just permissible, but positively right, to eat these animals whose comforts depend upon our doing so." This argument is not totally dissimilar from the one made by large corporations who hire children as laborers in poor countries and say that if they do not exploit them, the children will not find work at all. It may be true, but it is no justification for exploitation. The bottom line is the motive in both examples. Employers want to gain the maximum profits with the least expenditure, and the same is true of just about any farmer. Granted, some are more humane than others; all, however, are vulnerable to market conditions. Those pig farmers, and there are some, who insist on raising their pigs outdoors, to give them the semblance, at least, of a happier natural life, are soon run out of business by their more un-scrupulous colleagues. They simply cannot compete with somebody who has no restrictions on what he will do to turn a profit. The problem in both cases, though, is that the farmers are thinking about themselves foremost. The pigs are, at the most, secondary. If they could be kept happy with minimal expense, fine. If not, so be it. This thinking affects what one considers to be humane treatment; one is always lowering the bar driven by forces one does not really under-stand.

At the piggery, Joyce de Silva and I stood entranced at the stalls of the sows who had just given birth. There is something wondrous about watching six-inch-long squirming piglets pushing at their mothers and sucking voraciously. How, we wondered aloud, could people bring themselves to eat these adorable animals? The farmer acknowledged that everybody seemed to like watching small piglets. But people also like the taste of baby pigs. Indeed, they do. Professor Scruton justifies that as well: "Young animals have been slaughtered without compunction from the beginning of history."[36] (Things done "without compunction" are exactly the problem. Needless to say, Scruton's is a dangerous argument, for it could justify infanticide, racism, child abuse, slavery, and many other practices that go far back in human history but which most people would like to see abolished, despite their venerable genealogy.)

I left that piggery convinced that there was only one place for pigs, short of where nature intended them to live, free in their own world, and that is a sanctuary, a place where they could live for the rest of their natural lives without fear of being slaughtered for food.

Recently, I had the opportunity to roam the many acres of land given over to formerly abused pigs of all kinds—wild boars, Vietnamese potbellied pigs, and pigs saved at the last minute from slaughter—in the sanctuary known as "Pigs" in West Virginia, and to see the many nests that the pigs build for themselves. In the wild, female sows getting ready to give birth will often construct protective nests as high as three feet. They line these farrowing nests with mouthfuls of grass and sometimes even manage to construct a roof made of sticks—a safe and comfortable home-like structure. On modern pig farms, where the mother is forced to give birth on concrete floors, her babies are often crushed when she rolls over. This never happens in the wild because the baby simply slips through the nest and finds her way back to her own teat. When born, wild pig babies are striped for the purpose of camouflage, the shadings of their fur resembling the shadowy light of the forest. When the piglets arrive, they are precocious in one respect: they can move about and fight within minutes of being born. In all other respects, however, they are dependent on the nest, where they remain for at least ten days, since they are prone to hypothermia and can die of hunger quickly. The sharp incisors and canines with which the piglets are born are appropriately called needle teeth. The general consensus is that they use both their teeth and their motor skills to reach the most desirable of the sow's teats, the ones closest to her head, where they are in less danger of being knocked off by movements of her hind feet. Within forty-eight hours of birth the litter establishes a "teat order," and from then on each piglet suckles only that particular teat on his mother. This seems to be a form of dominance hierarchy, or at least has always been so interpreted. If allowed, piglets suckle for eleven to thirteen weeks or even longer. On factory farms, however, they are usually removed from the mother at three to four weeks and never see her again.[37] This is so that, having stopped suck-ling, the mother can be returned into the breeding cycle and be made pregnant again a week or two later. The thought is for profit, never for the pig or her young.

Piglets in a single litter can belong to different fathers, since do-mestic sows have a relatively long estrus period of two to three days and mate readily with different males, resulting in litters of mixed paternity. Evidently, unlike other ungulates, the males do not form harems or guard the female from the approach of other males after coitus. I have not heard of male pigs killing infant pigs, and perhaps this is because they cannot be certain who their children are and take no chances.

Wherever I walked in the sanctuary, animals walked with me. There is a peculiar and particular pleasure in walking in a natural setting accompanied by the true owners of that environment. If this gave me a thrill comparable to no other, it also seemed to please them. What on earth was going on in their minds as they walked slowly along? Wild boars are usually notoriously wary of humans, for good reason, having been hunted almost to extinction in some parts of the world and despised and feared just about everywhere else. Yet here I was walking in a field filled with domestic hogs, Vietnamese potbellied pigs, and wild boars. One wild boar spent the entire time I wandered the many acres the animals have at their disposal walking right next to my leg. What was the appeal for that boar who clung to me? Was it sheer novelty? The creature had every reason to loathe and fear my kind, and me, yet clearly he did not. It seemed that he wanted me to rub his tummy. Sure enough, the boar rolled over, all six hundred pounds of him, and looked positively ecstatic as I warily stroked his stomach. I say warily because he was formidably armed with strong, sharp tusks, and I know that even elephants have a healthy respect for the fighting ability of a full-grown wild boar. Yet this one had nothing more sinister on his mind than getting his full quota of tummy rubs.

Pigs, Kim Sturla has told me, are extremely tactile and love to be touched more than anything else. The minute you touch them, they close their eyes and wait for a massage. It gives them enormous pleasure. It occurred to me that if every person who ever ate bacon could experience the peacefulness and mutual trust of this encounter, if they could observe the majesty of these animals in their own environment, the pork industry might disappear.

At "Pigs," I was surprised to find how curious these pigs were. I was there to observe them and their habits, but they were equally interested in observing my habits and me. They followed me; they walked up to me with confidence, smelled me, and often nudged me to see what I would do. Clearly these pigs were aware that I meant them no harm and were doing anything but avoiding me. They would often approach me, push me roughly with their noses near the back of my knees, eager for me to scratch their backs or their tummies. Given that each and every pig there had a history of abuse of one kind or another, this spoke of a remarkable ability to forgive and forget. Here was yet another way in which, to my mind, pigs and dogs resemble one another. Rare is the dog not capable of forgiveness. To remember is human, to forgive porcine or canine.

Yet pigs have notoriously good memories. The author Sy Montgomery told me about her pig, Chris. A year after the vet first visited Chris to give a vaccination, the vet returned, and Chris started to scream at the top of his lungs at the first sight of the man. He also recognized and issued love grunts to a former next-door neighbor, a teenaged girl who returned to visit him after more than a year's absence, looking and probably smelling quite different from how she had the year before. People who work with pigs acknowledge this recognition. Stanley Curtis, professor of animal sciences at Pennsylvania State University, who says that the purpose of his research is "to have a conversation with a pig, to know how they are feeling,"[38] speaks of how pigs can remember people from three years earlier. Since we know wild pigs are capable of recollecting earlier experiences, would it not stand to reason that they can also remember the emotions they felt at the time? And if they can recollect, so might they be able to anticipate. Can they imagine their future? If they have any knowledge of their end, this must be terrifying.

The pig's capacity for pure pleasure is often remarked upon. Sy Montgomery tells me that what draws people to her pig Chris "is the chance to witness—when he eats, when he sleeps outside in the sun, when he is grunting with pleasure as you stroke him—pure joy. Chris exhibits this wonderful, unadulterated, pig-greedy joy. When he is happy, he is utterly caught up in savoring the sensual pleasures of life. He is utterly absorbed in his food. He exhibits great concentration, like you would see on the face of an artist at work. When he suns, he closes his eyes and gives himself over to pleasure. And when you pet him, he gives out these great heartfelt love grunts." Evidently these sounds are often made at an infrasonic level, which means that humans cannot hear them, but it allows pigs, like elephants, to communicate with one another without alerting a predator to their presence.

Louise van der Merwe, the founder of the magazine *Animal Voice* in South Africa, witnessed a full-grown pig gambol in frenzied delight:

"We let these boars out for 10 minutes every day to keep them healthy," the farmer told me.

"Really?" My face lightened. "Do they enjoy it?"

The farmer asked a nearby laborer to let one of the boars out while we went to wait outside. The boar's big body emerged from the shed door and he trotted heavily on his short legs along a narrow cement passageway leading to an enclosed strip of sand that ran along the back of the shed.

As his front trotters reached the sand, he suddenly broke into a frenzy of excitement, maneuvering his big, bulky body back and forth and up and down like a bucking bronco. He stopped momentarily to dig his snout as deep as possible into the sand, and then began to frolic and gambol once more. These pigs were simply expressing natural, normal pig behavior, acting as pigs have acted for the millions of years before humans deprived them of their birthright.

I have been told by many people who live with pigs in farm sanctuaries that they notice, on a daily basis, many common porcine emotions: contentment, happiness, love, grief, fear, anger, sorrow. Some are convinced that pigs have a concept of death and have feelings about it, much like elephants. I heard about a pig who, at three years old, lay dying from salt poisoning. To ease his suffering, an anesthesia was injected into his muscle, and then a euthanasia solution in his vein. He lived in a pen with fifteen other pigs. After he died, the pig was dragged to a fenced-in burial site. His companions walked alongside until they reached the fence, where the fifteen pigs lined up and watched as their friend was lowered into his grave.

They made what can only be described as a strange moaning sound. To those listening, it sounded like they were saying good-bye.

Piglet, featured in the Preface, is apparently not the only pig known to have sung to a full moon. They do seem to enjoy the moonlight. They are, after all, nocturnal creatures in their natural life. Several people who live near pigs have told me that they have awakened to strange sounds during the night. Sure enough, out their window, they spotted a pig looking up to the full moon, emitting mournful sounds much like singing. Are they singing? Perhaps these mournful sounds comprise a song of lamentation at their earthly existences.

Again and again, I hear stories about the remarkable emotional sensitivity of pigs. Gene Bauston, from Farm Sanctuary, an animal sanctuary near Sacramento, California, told me about the friendship of Hope and Johnny, a story that illuminates pigs' sense of loyalty and how they can feel emotionally connected. Hope was rescued from a stockyard with a severely injured leg. Nothing could be done for the leg, and she had to learn to live with greatly restricted mobility. Able to scoot around the barn on her three good legs, she could not walk. Johnny, who was much younger than Hope, bonded closely with her. At night, he would always sleep right next to her, keeping her warm on cold nights. In the morning, when Bauston would bring Hope bowls of food and water, Johnny would stay with her to keep other pigs from interfering with her or taking her food. During the day, Johnny would spend most of his time just hanging out in the barn with Hope. When Hope died of old age, Johnny was still a young and healthy pig. Maybe he knew nothing of death. The death of his closest friend seemed to devastate him; he died suddenly and unexpectedly within a couple weeks after Hope, perhaps of a broken heart.

There are cultures in the world that revere their pigs. In Papua, New Guinea, villagers have a special relationship with pigs. Pigs, say the Enga men of the western highlands, are our hearts.[39] Highlanders in the Nondugl area in the Middle Waghi on the same island, insist that pigs are so central to their social and religious life that no amount of money could ever persuade them to sell one. The piglets among the Siuai of Bougainville's Great Buin Plain on the same island, share their owners' food, are baptized, and are each given a ritual name.[40] When ill, they receive special magical medications and the women chew up tubers for them to eat. While the men "own" the pigs, it is the women who take care of them, allowing them to share their sleeping quarters, even nursing orphaned piglets.[41]

Those who live with pigs often speak of them as we normally speak of dogs—intelligent, loyal, and above all, affectionate. Each one, I am continually reminded by people who know them, is a complete individual, like no other pig. Recently, a pig sanctuary helped to save the life of a Vietnamese potbellied pig from Kentucky who had been severely abused and was not able to use her back legs. Pigs will bond with humans, with dogs, with turkeys, and other animals. This pig chose to live with the chickens rather than the other pigs. No doubt her disability had something to do with preferring the unthreatening companionship of chickens to the more aggressive boars, but for all we know she simply liked their company. The point is that there was nothing automatic or inevitable in her choice; it was something that grew out of her earlier experiences and her personality, just as it would for a human being.

The capacity, even talent, for cross-species friendships that pigs possess is not confined to the domesticated pig. Wild boars also clearly possess this quality. In his remarkable 1986 book, *Mein Leben unter Wildschweinen* (My Life with Wild Pigs), Heinz Meynhardt writes of his acceptance into a herd of wild boars.[42] Were Meynhardt to have seen these boars only as quarry to be hunted, his detailed knowledge about their social relations, their child rearing, and other intimate aspects of their lives could never have been compiled. Extraordinarily sensitive to danger as boars must be to have survived centuries of active hunting, why they would allow a member of the enemy into their ranks is mysterious and humbling.

Somehow they have generalized concern and curiosity for their own kind into an equally strong attachment to us, their ultimate predator. In their emotional generosity, which seems to be a characteristic of all pigs everywhere, do they allow for our moral weakness, forgiving us in a gesture of compassion which we would do well to emulate? These are remarkable creatures, worthy of our respect and deserving of our deepest apologies. It is time that we turn to pigs not as animals to be slaughtered for our table, but as distant family who have some special and deep affinity with us and are only waiting for a signal that we are at last ready to live with them in some kind of equality before they reveal to us with porcine exuberance the whole range of their complex emotional selves.

TWO

Does the Chicken Need a Reason?

A chicken flew into my arms. I didn't even know that chickens could fly, and suddenly one was landing on me. It happened when I was visiting a farm sanctuary. If I had been younger I would have asked my parents if I could take her home, please! After all, she *chose* me. Never mind that she chose everybody; she was a particularly friendly chicken. She made soft, strange cooing sounds and nestled into my arms like a happy kitten. I was won over. This was no ordinary chicken, I decided.

In fact she was an ordinary chicken, but simply one who had no reason to believe that people were after her. This is how chickens and humans would relate to one another if one was not exploited and the other doing the exploiting. Very much like cats and dogs. They just wait for the chance.

When I tell this story to people, they look bewildered. "But it's just a chicken." (They seem to feel uncomfortable using "he" or "she" instead of the impersonal "it" to refer to a chicken, as if giving chickens a gender would make them too personal or too real.) There is a beautiful and popular book called *The Fairest Fowl*, which contains photographs by Tamara Staples of championship chickens. When people see personality and emotion in the photos, they are accused of reading human emotions into chickens: "I think I begin to understand why the people who breed birds have no interest in photos that show chickens' true personalities," writes Ira Glass in the book. Chickens, he says, "may be capable of affection or loyalty or maybe even pride, but if so, they feel these feelings in an ancient and bird-like way, like glassy-eyed visitors from another world."[43] It is true that chickens are visitors from another world, but that world lies just around the corner, just a hair-breadth from human contempt. If we could drop our arrogance for just a moment, we might gain a glimpse of this other, mysterious yet enchanting world.

We read some animals so easily. Dogs, for example. We know dogs like having us around; they tell us so in ways that we understand. Charles Darwin recognized that dogs have at least four different barks to convey their feelings to us, but dogs are not the only animal species that talk. In *The Descent of Man*, Darwin reports, quoting the biologist Jean Charles Houzeau, that "the domestic fowl utters at least a dozen significant sounds."[44] Significant, Darwin would say, to each other. It would perhaps have been too great a logical leap for Darwin to wonder if chickens, like dogs, might be attempting to communicate with us while we have not been listening to them.[45] We take it for granted that dogs want to be part of our lives; but is it not also possible that the same is true of chickens and that we have not heard their requests because we have failed to understand their language? Darwin was impressed by the sense of humor displayed by dogs when they tease us, urging us to approach them with a stick lying nearby. When we come close, they run off with the stick and wait for us to try again to get hold of it, their tails wagging with delight in the game they are playing with us. Chickens, some people say, will never play with us. But how do we know this if we have never given them the opportunity? How many people have ever treated a chicken as they do a dog?

In fact, if we are to believe William Grimes, the restaurant critic for the *New York Times*, who has written a delightful book about an intense relationship he had with a large black Australorp chicken who took up residence one day in his backyard in Astoria, Queens, some chickens actually have a highly developed sense of humor. He writes that this particular Chicken (the name he gave her) seemed to have had an appetite for play: "Was it pure coincidence that she liked to sneak up on Yowzer, the cat most likely to develop a nervous twitch when caught unawares? Time after time I saw the Chicken trot up delicately when Yowzer had his back turned, squawk a couple of times, and then watch as the cat leaped a couple of vertical feet. The Chicken, after a successful ambush, would run off jauntily, with a cackle that sounded suspiciously like a chuckle." [46]

Perhaps if we had realized they are birds, with all the wonderful characteristics of birds, we would have paid closer attention to the ways in which chickens can enchant us. Take dust-bathing, for example. We call it a bath because the chicken finds a small indentation of dry earth and then proceeds to immerse herself in it as into a warm bath. The earth cleans her feathers. The first time I saw a chicken taking a dust bath, stretching out one iridescent wing and holding it up to the sunshine, then settling into the warmth of the afternoon only to fly effortlessly to a tree to roost in the evening, I was astonished. I did not know a chicken could fly into a tree. My surprise was a product of pure ignorance. I simply did not know chickens, nor did I know the many people who are passionately interested in them and in their well-being. I am entirely to blame for this ignorance. Least of all can I hold the chicken herself responsible!

Karen Davis is one of the foremost authorities on the lives of chickens and the founder of United Poultry Concerns, an organization in the United States dedicated to promoting knowledge about them. She tells me that chickens confined their entire lives will still perform "vacuum" dust baths in wire cages, so strong is their instinct to keep themselves clean. It is an empty gesture, for they can only behave *as if* they were outside, with real dirt to revel in. And of course the minute they are let out into a natural environment with real dirt, they will immediately dust-bathe. If you go to a farm where chickens are allowed to roam free, you will see them taking such baths frequently. It cleans their feathers, removes parasites, and gives them enormous pleasure.

Chickens have been with humans from almost the beginning of domestication. Like pigs, they are among the few domestic animals that can be found worldwide. According to F. E. Zeuner it was the "old-fashioned, multicoloured cock, hardly more than a large edition of the cock of the red jungle fowl which abounds in the woods of northern India" that was first domesticated in the Indus Valley civilization for the purpose of sport. In Greece, they were bred for fighting games, so popular that the Athenian government organized them. Zeuner notes, "It was accepted in its Eastern capacity as a divine symbol of light and health." As a symbol of fertility, the hen has always been preeminent, because of her prolific laying of eggs. Chickens, in ancient times, were rarely eaten. There were religious taboos in most places forbidding it; not however, in Greece. Rather these wonderful birds were kept for eggs and fighting alone. Or perhaps, as a timepiece, for as Zeuner points out, we are apt to "forget how much the ancient farmer relied on his cock to get him started in the morning." [47]

There are no places on earth, it would seem, devoid of chickens. Wherever there are humans, there are likely to be chickens. Yet while we have shared our lives with chickens for thousands of years, we understand them most imperfectly. I can't see much benefit in this relationship for chickens.

For example, why should they have become subject to that all-purpose insult—birdbrain? It is true that unlike mammals, birds do not have a neocortex, the region of the brain believed to be the seat of higher mental processing. But this does not preclude pigeons, for example, from an astounding ability to solve mental rotation problems when they are tested with comparison shapes rotated at various angles relative to a sample shape. Chickens, as far as I know, are never used for these tests, but humans tested on exactly the same task make more errors and need longer to react. Flying and looking down at objects gives pigeons proficiency in something in which humans are deficient. The scientist who performed these experiments concluded that pigeons were geniuses in comparison with humans! [48] "Different" need not mean less, and we should be cautious about comparing the intelligence of different species. Professor Lesley Rogers, who holds a chair in the Department of Physiology at the University of New England in Australia, ends a book about the brain and behavior of the chicken with these words: "In my opinion, there is a demand to understand the cognitive abilities of the domestic chicken above all avian species, because this bird is the one we have singled out for intensive farming. *Gallus gallus domesticus* is indeed the avian species most exploited and least respected." [49]

Whole books—of such importance that they have changed human understanding—have been written about earthworms (Darwin), bees (von Frisch), ants (Wilson), and other seemingly minor fauna. A comparable work for chickens does not exist. The American historian Page Smith and the biologist Charles Daniel have gone so far as to say that the brain of a chicken "represents to even the most capable and sanguine neurophysiologist a structure of almost unimaginable complexity."[50] Why should we not accord chickens intellectual respectability?

David Premack, the psychologist who worked on ape language, argued that even if chickens had a grammar, they would have nothing interesting to say.[51] Interesting to whom, one wonders. Despite this cavalier attitude, recent research seems to indicate that chickens have important things to say to one another, critical things that can mean the difference between life and death. Not only that, there also appears to be what scientists call an "audience effect" in chicken calls. What sound a chicken makes entirely depends on who the audience is. A rooster makes one sound if he wants to tell a hen about food and quite another one when he wants to alert the whole brood to the danger of a looming predator. He is not, as previously thought, merely making random vocalizations; he is communicating essential information. He is, in effect, talking. Roosters are well aware to whom they are addressing their sounds, and at last scientists are recognizing an increasingly wide range of verbal calls. The sounds have been there all along; what has been missing is our knowledge. In fact, this has become a hot area of scientific research, with evolutionary biologists like Peter Marler at the University of California at Davis actively engaged in furthering our knowledge. For instance, a rooster is capable of indicating to the hen the quality of the food he tastes through the type of call he makes. Moreover, the rooster seems capable of deception. In order to call the hen to his side when he reckons that she has ventured too far away, he will use a food call, even though there may be no food present. If this "is deliberate deception it provides evidence for complex cognitive ability not previously recognized in the chicken," writes the leading researcher on the brain of the chicken, Lesley Rogers.[52]

Karen Davis, in her excellent book *Prisoned Chickens Poisoned Eggs*, points out that each rooster can recognize the crow of at least thirty other roosters:

> If a rooster spots danger, he sends up a shrill cry. The other roosters echo the cry. Thereupon, the whole flock will often start up a loud, incessant, drum-beating chorus with all members facing the direction of the first alarm, or scattering for cover in the opposite direction. When it looks safe again, an "all clear?" query goes out from the rooster, first one, followed by the others, in their various new places. Eventually the bird who first raised the alarm sends up the "all clear" crow, and a series of locator crows confirms where every other rooster and his sub-flock are at this point.[53]

The calls are almost identical to those of their wild ancestors, conveying, according to Valerie Porter, a contemporary English authority on domestic fowl, "food discoveries, alarms, territorial claims, concern, fear, pleasure, frustration, dominance, appeasement and so on." The wild cock and the domestic rooster both crow at dawn but also toward the end of the day, clearly their way of calling the hens to roost in the trees, where they are safe from predators. When they find something that is delicious to eat, they will call their favorite hen, crooning to her in a special voice reserved for just this occasion. The female in turn does the same to draw the attention of her chicks to a particular food item. You can see a rooster picking up a choice morsel, then putting it down again, and repeating this until the hen, duly called, takes it from him.

Humans and chickens are both social species that respond to social cues. People have been watching chickens in their backyards and in other domesticated situations for thousands of years but have failed to observe their remarkable capacities because the only close contact they have had with them was in conditions outside the birds' natural social setting. In other words, only if we join a flock of chickens, or allow chickens to join our households, will we have the opportunity to see and test their potential for equally remarkable levels of intelligence and emotional interaction.

Exactly the same problem happened with parrots. Zdenek Veselovsky, the director of the Prague Zoo in the 1960s, wrote, "No parrot can learn to connect two or more words in order to reach a desired goal . . . this ability is typical of our species alone."[54] He was wrong. In her recent book about Alex, the African gray parrot who understands what he is saying, Irene Pepperberg complains that "for too long animals in general, and birds in particular, have been denigrated and treated merely as creatures of instinct rather than as sentient beings."[55] Dr. Pepperberg does not think that Alex is some kind of genius just because he can indicate color, shape, number (including what "none" means), whether an object is the same or different from some other object, even abstractions and requests. All African grey parrots are capable of doing what Alex does. In fact, what applies to parrots may equally apply to the avian brain generally. At least in theory, chickens may be capable of the same intellectual achievements as Alex, even if their way of voicing them is more limited.

In the last few years we have learned that there may be more communication between the human fetus and the mother than was previously thought possible. We know that the fetus hears sounds in the womb; similarly, information is communicated by the embryos inside the egg to the incubating hen. Even before birth the chick is capable of making sounds both of distress and of pleasure, to which the mother hen reacts. A day or so before hatching, the chick often utters distress peeps. The mother hen then moves her body on the eggs or makes a reassuring call to the embryo, which is followed by a pleasure call on the part of the chick. In other words, the bond between the chicks and the mother hen starts before birth. This makes sense, for it allows us to understand why a chick responds immediately after birth only to the calls of his mother. He recognizes her voice. The researchers who first discovered this in 1983 concluded that we cannot know what is natural behavior for chickens if we examine only chickens who were hatched in an incubator. Those who did so simply failed to realize that a great deal of communication had already taken place between the mother and her chicks while they were still in the egg.[56] This is a warning to those who conduct experiments without taking into account what happens in a natural setting.

Deeply embedded in the chicken brain is the instinct to construct a nest to protect her young. This is really not all that different from the human parent's "instinct" to paint and furnish a room in anticipation of a new baby.

There is a lively description by Alice Walker of a mother hen she met in Bali that "is one of those moments that will be engraved on my brain forever." It was the first time, she explains, that she really *saw* a chicken: "She was small and gray, flecked with black, so were her chicks. She had a healthy red comb and quick, light-brown eyes. She was that proud, chunky chicken shape that makes one feel always that chickens, and hens especially, have personality and will. Her steps were neat and quick and authoritative, and though she never touched her chicks, it was obvious she was shepherding them along. She clucked impatiently when, our feet falling ever nearer, one of them especially self-absorbed and perhaps hard-headed, ceased to respond."[57]

The British philosopher Stephen Clark has pointed out that scientists rarely place quotation marks around words like "see" when used about animals, but are quick to declare words like "love" out of bounds. The complexity of the bond between a mother animal and her young is especially difficult for humans to study with objectivity because, in my opinion, it is startlingly clear that human mothers and animal mothers have so much in common. In *The Descent of Man*, Charles Darwin quotes the philosopher of science, William Whewell, asking "Who that reads the touching instances of maternal affection, related so often of the women of all nations, and of the females of all animals, can doubt that the principle of action is the same in the two cases?"[58] Darwin's great friend George John Romanes wrote that "It must be admitted, from what we know of hens, that the maternal feelings may be so strong as to lead to a readiness to incur danger or death rather than that the brood should do so."[59]

The mother hen—a phrase that has come to signify good mothering in humans—may appear to be doing nothing but feeding herself, at least to the naive observer. To the person trained to see what is truly going on, she is in fact imparting essential knowledge to her chicks. Christine Nicol and Stuart Pope from the Department of Farm Animal Science at the University of Bristol demonstrated this conclusively in 1996 when they gave hens unpalatable food, colored blue, which the hens learned to avoid.[60] What would happen when their chicks were brought in and were also given this unpalatable food, but were too young to know? Would their mothers intervene? No experiment of this kind had ever been attempted before. It turned out that the mother hens did respond and attempt to get their chicks to avoid the bad food and eat the good food by nudging them away from the bad food. They knew that what the chicks were eating was not good for them and were teaching them what to eat. The scientists involved said they were "sensitive to perceived chick error."

However, any such purposeful communication has been rendered irrelevant by Western factory farming mechanisms, which bring the chick its food by conveyor belt. Today, chickens are one of the fastest growing creatures on earth, genetically altered to grow twice as fast as normal—fast food on legs, bred to be fried and eaten within seven weeks of emerging from the egg. Others are permitted to grow into egg-laying machines, caged by the thousands in mighty sheds without a glimpse of the sun-dappled light of their natural habitat. Many people now go through life without seeing a hen in any other form than a corpse. Nearly a quarter of all commercially reared birds are lame and experience excruciating chronic pain. Scientists like the veterinary professor John Webster of the University of Bristol

School of Veterinary Medicine, who exposed this situation, have been accused of being speculative, or worse, anthropomorphic. But recently, an experiment was carried out where chickens were offered two different feeds, one with carprofen, an anti-inflammatory drug with analgesic properties, and another without it. The lame chickens preferred the food with the medication, showing "that lame broiler chickens are in pain and that this pain causes them distress from which they seek relief."[61] Have we abused our power? Are we indifferent to the suffering we create?

Implanted in our language is the knowledge that hens, when permitted to sit upon their fertile eggs until they hatch, are devoted mothers: "How often would I have gathered thy children together, even as a hen gathereth her chickens under her wings." (Matthew 23:37) Less well known, and not yet fully explored, is evidence of the altruism of the rooster. In his *History of Animals*, Aristotle drew the attention of ancient Greeks to a paternal quality in the rooster that, to this day, is controversial. "Some of the males have been seen before now, after the death of the female, busying themselves about the chicks, leading them around and rearing them, with the result that they neither crow any more nor attempt to tread."[62] How real is this fatherly love? The sixteenth-century Italian writer Aldrovandi took up where Aristotle left off, and in his case there can be no doubt that he is writing from direct observation:

> He . . . is for us the example of the best and truest father of
> a family. For he not only presents himself as a vigilant

guardian of his little ones, and in the morning, at the proper time, invites us to our daily labor; but he sallies forth as the first, not only with his crowing, by which he shows what must be done, but he sweeps everything, explores and spies out everything. [When he has found some food], he calls both hens and chicks together to eat it while he stands like a father and host at a banquet . . . inviting them to the feast, exercised by a single care, that they should have something to eat. Meanwhile he scurries about to find something nearby, and when he has found it, he calls his family again in a loud voice. They run to the spot. He stretches himself up, looks around for any danger that may be near, runs about the entire poultry yard, here and there plucking up a grain or two for himself without ceasing to invite the others to follow him . . . to these characteristics add . . . the fact that he fights for his dear wives and little pledges to fortune against serpents, kites, weasels, and other beast of the sort and invites us to a similar combat whenever the occasion present itself.

Who was this man who writes with such affection about chickens? Ulisse Aldrovandi was professor of natural history at the University of Bologna. He spent decades researching a major work about chickens and clearly loved this bird.[63] Not content merely to study chickens, he lived with them. At his country home he "raised a hen who, in addition to the fact that she wandered the whole day alone through the house without the company of other hens, would not go to sleep at night anywhere except near me among my books, and those the larger ones, although sometimes when she was driven away she wished to lie upon her back."

Valerie Porter, who has written extensively on domestic fowl, notes that "quietly stroking a caught-up fowl can calm the bird into an almost hypnotic state, and laying the bird on its back also seems to quieten it."[64] This may be due to the bird feeling trapped and adopting a last-minute survival technique, much like a rabbit who freezes when caught. It may also be a purely physiological response to being put into an unaccustomed posture, like chickens carried to market held upside down by their legs, not calm, but petrified.[65] Why, though, would Aldrovandi's hen seek out such dangerous positions? One of Freud's more brilliant disciples, Otto Fenichel, invented the term "counterphobia" to explain why some people seek out the very thing that most frightens them: those most afraid of heights "choose" to become mountain climbers. It is a way of attempting to master the fear. Could Aldrovandi's hen have adopted a counterphobic night ritual?

Why would a particular chicken forego the company of her own kind and choose to spend the night with a human? Was she imprinted? Or can we say that if Aldrovandi could feel a particular pleasure in the company of his hen, such that he allowed her the freedom of his house, is it no less possible for the hen to reciprocate these feelings?

Imprinting may account for the success with which wild jungle fowl have been turned into domestic chickens. Juliet Clutton-Brock, an expert on the history of domestication, points out that the flight distance—how close an animal will let you come before flying off—is a more or less fixed length for each group of animals, and that when humans were Paleolithic hunters in the northern tundra, the flight distance between human hunters and their potential prey was quite short. In other words, before guns were invented, we could come quite close to birds. It was not unusual, even in those prehistoric days, for hunters to play with and nurse the young of any animal they found. If a wild jungle fowl were to abandon her eggs in the vicinity of human habitation, the chicks emerging from the eggs could perhaps have seen a human and would be genetically programmed to believe they had found their mother. It would take a particularly hard-hearted person to walk away. An already well-developed instinct in humans to share inclined at least some women to feed the young as well.[66]

Konrad Lorenz, who won the Nobel Prize and is considered the founder of ethology, is usually credited with discovering imprinting in 1935. In fact he simply named it. The phenomenon itself has always been present and was certainly known to our Neolithic ancestors.

In his book *Illumination in the Flatwoods*, Joe Hutto uses his experience with wild turkeys to give us a most persuasive description of imprinting. Nearly two dozen eggs were hatched in his presence, and he became the leader of a flock of extraordinary birds. He lived among them for one year in the forests of Florida and learned more about these birds than anyone else has ever recorded. Hutto discovered that they are extremely intelligent, not at all the creatures of human myth. Of course, turkeys do not drown in rain, as a silly legend has it; on the contrary the position they take—head up, neck raised, body erect, and tail down—keeps them relatively dry by exposing as little as possible to the rain.[67] He tells us "I have never kept better company or known more fulfilling companionship." Hutto is driven, in spite of his scientist self, to recognize that "in the most fundamental sense our similarities are greater than our differences." He considers himself privileged to be in their presence, feeling less desolate, less isolated, as he is "bathed in the warm glow of these extraordinary creatures." For a naturalist who has hunted wild turkeys all his life, he goes about as far as anyone can in his affirmation of their uniqueness: "As we leave the confines of my language and culture, these graceful creatures become in every way my superiors. More alert, sensitive, and aware, they are vastly more conscious than I. They are in many ways, in fact, simply more intelligent. Theirs is an intricate aptitude, a clear distillation of purpose and design that is beyond my ability to comprehend." He describes his friendship with one particular bird, Turkey Boy:

> Each time I joined him, he greeted me with his happy dance, a brief joyful display of ducking and dodging, with wings outstretched and a frisky shake of the head like a dog with water in his ears. Occasionally, he would jump at me and touch me lightly with his feet. His anticipation and enthusiasm made it difficult for me to disappoint him.[68]

What draws him to them, beyond their unusual intelligence, is "observing the absolute joy that these birds experience in their lives . . . they are in love with being alive."

If hens are the very model of female concern for children, roosters have been venerated for their ability to fight, as witnessed by the almost universal appeal of cockfighting to men in cultures ranging from Bali to ancient Greece, Tudor England, and other parts of Europe and the Far East.[69] In the United States, cockfighting is legal in

New Mexico and Louisiana, and carried on illegally in many other states. In fact, it has even been suggested that chickens were domesticated only because men liked to see the roosters fight. In a natural setting, roosters may fight, briefly and often ritualistically, to defend a territory, or a female. It is, however, highly unlikely that wild and feral roosters would, under natural conditions, ever fight to the death. It is so much easier to simply fly away. Humans want to see blood, not roosters.

(Might it have been the alarm-clock function of the rooster that led to domestication? In *Chanticleer*, Edmond Rostand, the author of the ever-popular *Cyrano de Bergerac*, devoted an entire play to this theme at the beginning of the nineteenth century.)

To this day nobody is certain why the rooster crows at dawn—as well as three times during the night. The evening crowing seems to be an attempt to call the flock together to roost in safety in trees. In a posthumously published essay, Darwin saw this as a failure of instinct: "The cock-pheasant crows loudly, as everyone may hear, when going to roost, and is thus betrayed to the poacher."[70] Of course, poachers were not part of the evolutionary history of wild fowl, and so they could not be expected to be alert to them in the past. Darwin continues: "The wild Hen of India, as I am informed by Mr. Blyth, chuckles like her domesticated offspring, when she has laid an egg; and the natives thus discover her nest." Can this also be considered a failure of instinct, when the wild hen so rarely had encountered a human in her territory, and had only done so relatively recently in her history?

As for crowing early in the morning (at least forty-five minutes before what we consider to be dawn), it must be said that, with his superior vision, a rooster perceives the light long before a human can. Many species of birds, including chickens, can see well into the infrared spectrum, something that allows them to detect polarized light, and therefore direction, with greater sensitivity than humans. They see into the ultraviolet spectrum as well. Many birds are also able to hear and discriminate low-frequency sounds far below the human range of detection. This, it is now thought, is how migrating birds may be able to find their direction, perhaps using the sounds of ocean surf or wind passing around and through mountain ranges.

In 1865, Francis Galton, Darwin's cousin, better known as the founder of the sinister science of eugenics, wrote an influential article on domestication in which he put forward the hypothesis that animals were domesticated as a result of first being tamed.[71] He cites in that article a wonderful passage from Ulloa, "an ancient traveler," about South America:

> Though the Indian women breed fowl and other domestic animals in their cottages, they never eat them: and even conceive such a fondness for them, that they will not sell them, much less kill them with their own hands. So that if a stranger who is obliged to pass the night in one of their cottages, offers ever so much money for a fowl, they refuse to part with it, and he finds himself under the necessity of killing the fowl himself. At this his landlady shrieks, dissolves into tears, and wrings her hands, as if it had been an only son . . .

This is a scene that has been repeated over and over again on modern farms, especially where children are concerned. Perhaps recognizing a kinship with the birds, children—who are also subject to the whims of more powerful humans—sympathize with chickens and other animals on the farm and sometimes remember these first friendships and the traumatic event that normally ends them. When I was about eight years old, we lived next door to a man whose free-range chickens roamed his yard. He would ask me over to help him collect the eggs. I loved it. They were warm and perfectly shaped. Each egg struck me then (still does) as a kind of miracle, a poem of nature. But at the time I thought, *are we thieves?* Taking these eggs that belonged to the hen, not to us. The farmer assured me they were gifts, but I did not believe him. Especially when the eggs had blood in them. Once I even found a tiny chicken embryo. The hen clucked in what sounded like protest. Why did we take the eggs away from her and eat them? Was it a bit odd to eat the menstrual product of another animal? We unlearn this type of thinking fairly quickly and at a young age, but it haunted me a long time. I still do not feel right about eating eggs, even the eggs of free-range hens.

We forget that a valuable egg-laying hen can far outlive her capacity to lay eggs. Most farmers put profits before gratitude, and the same hen who has given thousands of eggs is slaughtered with hardly a second thought. The same neighbor who had had egg-laying hens by the dozen asked me to hold a particularly sweet chicken for him for a moment, while he went to get something. When he returned, he had an axe. I was six years old, and I can still remember the fear that gripped me as he seized the hen, sliced off her head, and threw it into the yard while, to my horror, the hen ran around spurting blood from her neck. I was sickened by the sight but even more perhaps by the ghoulish laugh our neighbor directed in my direction. After all, the hen had been useful to him, and so had I. What was to prevent him from deciding that I had outlived my usefulness? I never set foot in that yard again and don't think I would do so to this day.

The Dutch ethologist J. P. Kruijt stated that "no great differences exist in the behavior of domesticated and wild *Gallus gallus*."[72] Marian Stamp Dawkins, from the Department of Zoology at the University of Oxford, notes that, "Junglefowl, which are the wild ancestors of our domesticated chickens, spend long hours scratching away at the covering of leaves that hides one of their favorite foods—the minute seeds of bamboo. An ancestral memory of this way of life seems to have carried down the generations into the cages of our modern intensive farms so that even highly domesticated breeds have the same drive to scratch away to get their food—if they have the opportunity."[73] Science clearly recognizes here that there is hardly a difference in the behavioral need from the ancient free-living jungle fowl to our domestic hen.

In 1967, ornithologists Nicholas and Elsie Collias[74] made a discovery that many animal behaviorists would challenge, but which I am convinced is indisputable and highly significant: namely that these wild ancestors of our domestic chickens are keenly aware of being hunted and have learned to effectively use evasive techniques. They saw that when native beaters (who carry sticks and beat the bushes) tried to flush the birds, one cock flew high up into the branches, some sixty feet above the beaters, and watched them silently, without betraying himself by any sound or movement. He knew what the beaters were attempting to do. Eventually he gave a loud alarm cackle and the birds flew over the heads of the beaters.

Then he gave an all-clear call, a rallying call, and the birds reassembled. It was clear to the authors, highly respected authorities in their field, that wild jungle fowl are able to perceive the goals of human hunters and can attempt to evade them using deceptive techniques.

Recently I spent an afternoon with Juliet Clutton-Brock at her country home near Cambridge, England. She is the guardian of twenty chickens from six different breeds, all living the perfect life. What a pleasure it was for me to see these sleek, healthy, colorful hens and roosters strut happily in the sunshine, stopping from time to time for a quick and ecstatic dust bath. When I asked if she had ever observed depression in chickens, she told me somewhat firmly: "My chickens are *never* depressed." Indeed, they are free to roam the one-acre lawn and gardens during the day, and in the evening, they put themselves to bed in their own fox-proof roost. They do this themselves because, as Clutton-Brock explained to me, "Chickens always know exactly what the time is, and what should happen at all times." She has been amazed to see one of her hens, who is five years old, suddenly, in this her fifth year, become increasingly tame and increasingly friendly. Whenever Clutton-Brock returns to the house from the garden, the hen simply follows her into the house and moves with her from room to room, curious about everything she does. It has been such a sudden change that one cannot help wondering about the capacity of chickens to adapt to new situations. This chicken has obviously decided that it is safe to be friendly. Every chicken has an individual personality, she reminded me. As with pigs, this is something I hear over and over again from people who have lived in close contact with chickens. Some are utterly indifferent to humans; others are completely obsessed with us and wish only to be near someone whom they consider a close friend.

People who live with chickens say that they are naturally sociable with each other and will also gather around a human companion and stand there serenely preening themselves or sit quietly on the ground beside someone they trust.[75] When I visited Karen Davis, I saw this for myself. As we sat in the garden talking and I picked her formidable mind about chickens, a group of them slowly and quietly gathered around us, just for the company it would seem. It may be surprising to think of a chicken showing trust, but it is a decision that must be made on any given day, whether to trust a specific person or other animal or not. Instinct does not help here, for the chicken's instinct is not to trust anyone who can be considered a predator. Wild jungle fowl are so notoriously shy and elusive that it is difficult even to catch a glimpse of them. Nicholas and Elsie Collias, who were among the first ornithologists to study them, report that even people with a lifetime of experience in the forest rarely, if ever, have seen a bird caught by a predator, for "the red jungle fowl in nature is one of the wariest species of birds in the world".[76] The birds have learned the hard way that humans are not to be trusted; they mean them harm. To unlearn this requires a great deal of thought, based on experience.

"End-of-lay" is a terrifying British term signifying the end of a chicken's utility, as if a hen's only purpose were to lay eggs. In the United States, the term, equally nasty, is "spent hen." Clare Druce, the British founder of Chickens' Lib, an organization dedicated to liberating chickens from cruel and unusual punishment, rescued an "end-of-lay." The thin, featherless hen, who was going to be slaughtered, was brought by Druce in a total state of terror to an orchard. Such were her recuperative powers, Druce told me, that within twenty-four hours this hen, whom she called Felicity, was able to construct a perfectly formed nest. Her beak had been cut off, as were the beaks of all chickens in factory farms. (The practice has recently changed; the beak is cut off. This is for economic reasons, not out of consideration for the chickens.) Felicity picked up pieces of straw in her disfigured beak and carefully arranged the strands to build a protective shelter for her eggs. She almost disappeared from view in her deep, deep nest. Druce wonders whether she was making up for a lifetime of deprivation. In a short time, she had a new, healthy coat of feathers, her legs were normal, and she was happily enjoying the freedom to wander in an orchard in the company of a dozen or so other hens. Perhaps even more remarkable was Felicity's psychological recuperation, her ability to bounce back from a state of total deprivation to one where she not only trusted humans, but also derived obvious pleasure from being in their company. She died peacefully and was buried under a flowering cherry. How different from the fate of the billions of hens raised in factory farms the world over. I think we have to agree with the ancient philosophers, that animals know gratitude and recognize their friends from their actions.

Lauralee Blanchard lives on the island of Maui. In December of 2002, a factory farm was getting rid of "spent hens" for a dollar each if you bought four of them, bound together by the legs. People throw them into a large pot of boiling water and make chicken soup, she was told. So Lauralee decided to take home her four and give them a real life. One of the chickens was particularly affectionate and sweet. In spite of what she had seen of human behavior, she craved human company. Her leg was badly damaged by being tied too tightly when she was sold, yet she would limp over to Blanchard and with great difficulty jump up into her lap, settling down contentedly while being softly stroked. Many people have written me to tell me how loving they find chickens and what intense bonds can develop.

Karen Davis told me that her chicken Muffie became friends with her adopted turkey, Mila. As soon as they met, they liked each other and would go off into the garden looking for food together, and sometimes they would even delicately preen one another, smoothing their feathers and cooing softly. Davis said "one of their favorite rituals took place in the evenings when I changed their water and ran the hose into their bowls. Together, Muffie and Mila would follow the tiny rivulets along the ground, drinking as they went, Muffie darting and drinking like a brisk brown fairy, Mila dreamily swaying and sipping, piping her intermittent flute notes." They did not grow up together, yet they formed an intense bond. They did not mistake each other for members of their own species, but this did not seem to make a difference to their friendship. Clearly, they just liked each other.

Gilbert White, the great English naturalist, in his classic book *The Natural History of Selborne*, first published in 1789, wrote one of the earliest, most beautifully described accounts of a cross-species friendship, about a "very intelligent and observant person" who had a single horse, "and happened also on a time to have but one solitary hen. These two incongruous animals spent much of their time together in a lonely orchard, where they saw no creature but each other. By degrees an apparent regard began to take place between these two sequestered individuals. The fowl would approach the quadruped with notes of complacency, rubbing herself gently against his legs; while the horse would look down with satisfaction, and move with the greatest caution and circumspection, lest he should trample on his diminutive companion. Thus, by mutual good offices, each seemed to console the vacant hours of the other." [77]

Kim Sturla, of Animal Place, the sanctuary in California, told me about a hen she found at a city dump. This was an older chicken who had lost most of her upper beak and much of her lower beak during a botched debeaking. This painful "operation" is done so quickly, without, of course, anesthesia, that it is not at all unusual for it to result in horrendous injuries, which then go untreated since to do otherwise would be "uneconomic." She brought Mary, as she called her, home to the sanctuary, to live out her remaining years in safety. In spite of her deformation, Mary had remarkable confidence, a strong sense of self, as Sturla calls it. Mary became fast friends with Notorious Boy, a young rooster who was considered a gentleman, in contrast to the usual image of roosters. They spent all their time together, hardly interacting with the other birds. It was a kind of love, though not sexual. They would bask in the sun together, look for food and would always sleep close to one another. The spot they selected was far from the chicken barn. They chose to sleep on a picnic table outside the kitchen window. When the first winter rains came, and it began to pour, Sturla went outside to bring them indoors. She found them huddling close together, Notorious Boy's wing draped over Mary to protect her from the wind and rain, just as a mother hen would protect her chicks.

It is something of a cliché among animal behaviorists that wild animals do not tolerate disabilities, and that animals who are unfortunate enough to be born with a deformity or fall ill rarely last very long. I am dubious. Recent research on many species has shown that young animals born with serious disabilities are nevertheless able to live with the help of their mothers and sometimes other friends and relatives. This is particularly true of elephants but applies to many species. Indeed, animals may have no concept of "disability" in the way humans do. Inspiring in this instance is the account Kim Sturla gave of Helen, a completely tame hen who was found wandering the streets of San Francisco. She was totally blind, and dogs were mauling her when a homeless person took pity on the hen and rescued her. She was taken to the city's animal shelter, where a call was put through to Animal Place to see if they would be willing to give her a home. Helen was born with a condition called cryptophthalmos, meaning that her eyelids had never formed properly and therefore never opened. One foot was missing and one of her legs was several inches shorter than the other. Concerned on the first night that Helen might become the object of derision from the other hens and roosters, Sturla set up a special nest in the barn. But when she opened up the door the following morning, a triumphant Helen greeted her sitting proudly on the top perch. Blind and lame, she had somehow found this spot. Far from feeling derision for Helen's disabilities, the other birds stood in a kind of awe of her, and she lives to this day in complete harmony with the rest of the flock, preening her feathers, basking in the sun, dust-bathing with pure delight.

Many who write about animals have noted that chickens form

unusual friendships. Maurice Burton, in his book *Just Like an Animal*, tells of an aged hen, Aggie, who was almost totally blind and had become a pet wandering as she wished about the garden. Her owners could not pluck up the courage to put her down. She was protected by a bantam (a breed of miniature chicken) who became her inseparable companion, sunbathing and dust-bathing together. At night the bantam would lead Aggie to her roost. When Aggie died, the bantam went into a depression and within a week was also dead.

Consider the question often asked by scientists, including even those who are well disposed toward animals, as to whether the hen suffers from what she has never known. The Oxford researcher Marian Dawkins conducted experiments to determine what hens felt about their homes. Somewhat to her surprise, she found that hens who had been confined to battery cages, cages no larger than a sheet of newspaper, when given the choice between a small outside run with grass and the cages they had known all their lives, chose to stay in the cages. And fowl expert Valerie Porter points out that chickens taken from a battery cage "will be in a considerable state of what you might call cultural shock if they are deprived of the only type of environment they have ever known. In fact, they will cower in a corner in a state of petrified agoraphobia and it will take a great deal of time and patient understanding to rehabilitate them to real life."[78]

While Dawkins acknowledges that the more experience the hens had of the run, the more likely they were to choose it when they next had the choice, she still argues that finding hens "prefer being outside in a run to being inside in a battery cage does not tell us anything about whether they suffer in battery cages." Dawkins appears to lean toward the view that a chicken can be perfectly happy even if prevented from expressing her "natural" inclinations. Even in battery cages, she claims, "the animals might not, under any criteria, appear to be suffering."[79] I have to use her words here, because "might not appear" just makes no sense to me. I think she means that there is no way she can tell the chickens are suffering, though common sense tells us (and her, I imagine) that they are. Scientists believe that finding out how important certain things are to an animal is the only legitimate way to ask an animal what he or she feels. Are they right to make such an assumption?

We must use our own empathy—stretching it across the species barrier—as a means of knowledge if we wish to be freed from the sterile experiments of academia. I defy anyone to enter a shed with up to half a million chickens in it, spend an hour in that stench, and tell me that the chickens are happy, or that we cannot know whether they are or not. It flies in the face of common sense. One might as well argue that there is no such thing as "natural" behavior when it comes to humans, so wide is the range of our behavior. A child can be sat down in front of a television set or a computer for most of the day and could appear, from all outward signs, to be perfectly happy. An objective witness might still say that his happiness is limited and the sad effects of his obsessive interest may not be apparent to himself or others until years later. Richard Adams, the author of *Watership Down*, the great novel about rabbits, while no animal rights fan, nonetheless takes this more commonsense view:

It is plain enough that the hens—deprived of living space, of movement, of anything like a natural life-span—must suffer, but it is convenient and profitable to ignore this, and to try to stop the public thinking about it. "Well we all exploit animals, don't we?" "Hens are silly creatures." "What are you going to do for hens and eggs otherwise?" Nevertheless, consciousness of the infamy won't go away: most people have it ticking away inside them somewhere.[80]

If we don't know what a chicken wants, if we have never asked the question in all seriousness and been patient in our attempt to find an answer, we cannot know what they have been deprived of in the life we mete out to them. How much attention, for example, does a chicken require? When I visited Kite's Nest in the Cotswalds—the famed organic farm that is the model for Prince Charles's own farm—farmer Rosamund Young told me that recently a large group of visiting French agricultural students had gathered into a circle to learn about the farm rotation. The hens, feeling deliberately excluded, pushed into the center of the mêlée, stretched to make themselves as tall and noticeable as possible and tried to take part in the conversation, the only way they know, by singing loudly. Generally, to be ignored by humans is a good thing for chickens; it means they are not being exploited. But when they live in complete freedom, as at Kite's Nest, they want to be part of the life, they want to be acknowledged and included. In fact, Young tells me that in winter when she loads up her Range Rover with hay for the cows, the hens try their hardest to cadge a lift:

They know they are not supposed to, so they peck around the wheels nonchalantly and wait for an opportunity when our backs are turned. Then one will manage to jump in and hide amongst the hay. If the engine is running, we do not hear the triumphant singing and on occasions she will get away with it and not be discovered until the hay is being unloaded way up in the fields. Once I discovered one and lifted her into the front of the vehicle in case she fell out of the back and she stood on the seat and looked round her like a queen on an official drive-about.

It is strange to think that a chicken is a *bird*. This is because, with few exceptions (penguins and ostriches, for example), we tend to think of birds as flying creatures. People do not think of chickens as having the ability to fly. Chickens rarely fly. Having seen its wild ancestors, the Burmese fowl, also called the northern red jungle fowl, all over India and Bali, I can confirm that these birds fly, and quite well. Their evolutionary cousin, the eider duck, is one of the fastest flying of all birds. Not even swifts or swallows can outpace an eider, who might reach 60 mph in level flight, and has been described as "arguably the world's fastest bird."[81]

We tend to think of chickens living in the backyards of farms, enjoying the quiet life and the sunshine in the midst of their families, and out of gratitude, dropping eggs from time to time for human use. Alas, that is not how 99 percent of chickens live at all. They are incarcerated in small cages—each typically housing five hens in a space measuring eighteen inches by twenty inches and stacked three or five tiers high. The sloping wire floors cause severe damage to their feet and claws. There is no sunshine, the artificial light is kept dim, and the birds live in what can only be described as a form of hell. Some people are so incensed by this cruel practice that they slip into the facilities at night and liberate the animals from their cages. What is interesting is that they invariably report that within days the freed hens take to roosting in trees. They have retained an ancestral memory of what has given them pleasure (not to mention safety) over millions of years of evolution in much the same way that we seek out shade on a hot sunny day.

"Nonsense," a critic might argue, the birds do not *enjoy* roosting in trees, or bathing in the dust; "it is merely instinct at work." I wonder how we would respond if someone told us that we only loved our children because of some built-in mechanism or impulse to do so? We might well have such an inborn urge, but surely this only makes it easier for us to understand what we have in common with a hen. Moreover, the emotions we feel while obeying that instinct are still real, and surely it is those emotions that matter, not the source of them, and those emotions appear to be shared between humans and other animals, including the domestic hen. Lying in the sun, drinking water, sitting quietly in peace and contentment, the hen's feelings during these times are perhaps purer than they are with us, since they are unlikely to be contaminated by worries about the future.

When I was in Australia I visited Patty Mark at her home in Melbourne, where she rescues battery hens. The yard was filled with them. Mark's fearlessness is legendary: she will go to any lengths to protect birds who are being abused on poultry farms; for her it is a matter of moral duty. I have seen videos of Mark and her associates making their way to a vast shed containing almost a hundred thousand miserable chickens, starved of sunlight, fresh air, green grass, and blue sky. These intrepid chicken-saviors find their way inside and rescue some of the hens who are near death. She is justly proud of what she did, even if she had to go to jail as a consequence. It was said that she had stolen other people's "property," though she believes, and I agree with her, that the day will come when this word will never again be used in conjunction with a living being.

When I met Mark, it was a beautiful sunny day, and as I stretched out on the grass, with my then three-year-old son Ilan next to me, several hens approached to investigate. One in particular sat down next to Ilan and settled into what looked very much like sunbathing. When Mark showed me a video clip of this same hen in her former life, I found it hard to believe that an animal who had suffered so severely could have survived and shown such delight in close physical contact with the same class of beings who had been her tormentors. Mark and others who live with chickens claim on good grounds that chickens recognize certain people and have good memories for who has been kind to them and who has not. It would seem these hens showed a remarkable ability to forgive, or perhaps they were just able to discriminate.

We have attempted to crush the spirit of the domestic chicken, hoping the hen will not obey an instinct to roost in a tree. When she is in a cage with ten other birds, unable even to spread her wings, of course she cannot give expression to this instinct. But we have not succeeded in crushing her spirit. This we see the minute she manages to escape from her prison. In general, whenever chickens are allowed to revert to feral life, they reveal behavior that had not been seen or expected in the domestic chicken. What we have failed to see is therefore not because it does not exist but because the conditions we have created are so artificial that, instead of chickens, we are seeing in effect some kind of deranged bird, a sort of distorted version of the real chicken. Of course, as Karen Davis reminds me, they are no more artificial than are humans released from prison camps. They are living beings, infinitely more complex and interesting than any machine ever created, and unlike any machine now or probably ever, they *suffer*.

When the late professor David Wood-Gush and his colleagues released chickens on an uninhabited island off the coast of Scotland in the spring of 1975, they were surprised at what they found. While previous research on domestic chickens indicated that they are highly territorial birds, Wood-Gush found that "while the hens foraged no evidence was seen of any territoriality." Not only that, but the hens were perfectly at ease when the chicks of another hen entered their territory and became, however fleetingly, members of the family: "They often passed so close that their broods temporarily intermingled."[82] In a laboratory, a chick follows the hen, and there is nothing to be gleaned from this. But Wood-Gush was able to conclude "that the chick in the wild has a more positive relationship with the hen than one would expect from experiments on the following-reflex, as it is called, under laboratory conditions." And with aggression, too, the expectations from artificial conditions were reversed: The amount of antagonistic behavior seen in the adults in the non-breeding season was very small.[83]

A woman from New Zealand who lives with chickens, a civil servant by the name of Helen McNaught, became intrigued by my questions about their emotional lives, and sent me an interesting analysis:

> The first of our roosters was a handsome bantam with an insatiable sexual appetite that earned him the name of Randy.

The objects of his affections were our full-size brown shaver chickens, and they seemed delighted to receive his attentions despite the logistical difficulties inherent in the fact that he was approximately one-third their size. When we eventually separated Randy from the chickens, he went into a decline. You have never seen anything so depressed and miserable as an unhappy rooster. Or was I just reading emotion into the dejected droop of his once magnificent tail feathers, his loss of interest in food, his lack of attention to personal grooming? As soon as we let him back in with the hens, his tail feathers perked up, his appetite increased, and he regained his cocky macho strut, the king of the farmyard once again. He was clearly a much happier bird in his preferred environment.

I suggest, as a former psychoanalyst and someone concerned with the etiology of depression, that we would do well to examine depression in farm animals as a way of understanding human depression. In every case I have seen, the animals are depressed because they are deprived of their normal life. In the factory farming of chickens, natural instincts are frustrated, and this can only lead to unhappiness. Wild fowl, like all birds, do not lay a surplus of eggs, except in the spring when they are prepared to raise a brood of chicks. Chickens produce so many eggs not because it is natural for them to do so, but because light stimulates the hen's pituitary gland at the base of the brain, resulting in a greater amount of hormone, which in turn stimulates the ovaries of the hen. This research remained unknown until sometime shortly after the Second World War. If the lights are left on, hens will eat and lay for twenty-one out of twenty-four hours of the day, baffled by the continual disappearance of their eggs. Chickens now had to be put in a controlled environment in which sun, wind, rain, bugs, worms, sprouts, and growing things had no place. Chemicals must be added to their food to stimulate their appetite. Domestication, then, has not merely altered the natural behavior of the chicken, it has distorted it, twisted it into a perversion, a caricature of ordinary life.

Temple Grandin, an animal science professor famous for devising methods of killing cows "more humanely," in a paper presented to the National Institute of Animal Agriculture in April 2001, spoke of her disgust at seeing these factory farms for chickens: "When I visited a large egg layer operation and saw old hens that had reached the end of their productive life, I was horrified. Egg layers bred for maximum egg production and the most efficient feed conversion were nervous wrecks that had beaten off half their feathers by constant flapping against the cage. . . . Some egg producers got rid of old hens by suffocating them in plastic bags or Dumpsters. The more I learned about the egg industry the more disgusted I got. Some of the practices that had become 'normal' for this industry were overt cruelty. Bad had become normal. Egg producers had become desensitized to suffering."[84]

The conditions for chickens raised for meat are no better. I had wanted to see how broiler chickens are raised commercially for some time. Not easy to do. Such places are off-limits to the general public. Chicken suppliers do not want people to know the intimate details of how their cheap chicken comes to the dinner table. Recently, though, Tony—a friend of a friend of a friend—said he would let me visit his chicken farm, as long as I did not identify him with a last name or say exactly where the farm was. A few weeks ago, I drove to

Tony's. He took me to four shed-like barns secluded behind giant cypress shrubs, well out of view of the public.

"We are expected to keep them out of sight," he told me.

Nobody would ever guess what was in those gigantic green sheds. In each, 25,000 broiler chickens (the name refers to their fate, to be broiled for the table) are kept for approximately one month. The day they hatch, they are delivered by truck, swept into one of the giant barns, and live there for their one month of "life," until they are taken down the road to a slaughterhouse. As I walked in, I was almost blinded by the sight of 25,000 pure white chickens, packed up right against one another as far as my eyes could see. It was like a hall of mirrors, never-ending chickens, all the same size, lined up one next to the other, eating the food, drinking the water, in artificial light, and in almost total silence. I had expected noise and a terrible smell. But it was deathly quiet; the quiet, it felt to me, of despair, not contentment. Those closest looked up at me, and I had a horrible realization that I was letting these chickens down, even as I was there to understand and write about their plight in the hope that some people would see that killing them was wrong. But I would do no good for these 25,000 chickens. They would all be slaughtered as sure as I was standing there.

They were all eating; what else could they do there? The idea is for them to eat as much as possible, especially during the last week, so that they get to the right weight as soon as possible and with as little expense. Computers dole out the food, computers lift the feeding troughs, adjusting them to the correct height as the chickens grow; everything is looked after as befits a commodity. There is no sense that these are living beings. It would have been absurd for me to ask Tony if he thought they had any feelings. They are hardly seen as alive. Every day, Tony explained, he walks through this stiflingly packed room and picks up the dead and the dying chickens and disposes of them. He eyed me warily.

"You're not from one of those crazy animal rights groups, are you? Okay, then, well, I guess I can tell you, I also take out the ones that are not growing. It wouldn't pay, would it, to keep them there? No profit, they are just useless eaters." (The phrase resonated for me. "Useless eater" was used by the Nazis to describe the inmates of psychiatric institutions whom the Nazis wanted dead, and indeed did kill.)

Tony had to be very careful how he walked among the chickens, though. They could easily panic and then they would make a huge rush for the sides of the giant barn, and, well, "that many chickens rushing for the wall, a lot of them get suffocated." What panics them? "Oh, anything,"—(the specter of death, perhaps? The sight of their ultimate predator?)—"they are just, well, chicken you know. Ha ha."

Tony did not strike me as a cruel man. He was just going about his business. It was all only about money. He had four barns, so 100,000 chickens. Each was worth about 25 cents. So little? "Well, it adds up you know, 100,000 a month, year after year. But the barn with all its sophisticated computers and machines cost more than half a million dollars." No, he didn't particularly like chickens (except to eat), but he didn't believe in being cruel to them either. He knew plenty of other chicken farmers who had much larger barns, and far more of them, and some of them just did not care about the chickens at all, so the stench was horrible, and the conditions unbearable. Standing where I was, it was hard to imagine it could be much worse, but I knew Tony was correct.

When I looked at these chickens, I thought of Tom Regan, the preeminent philosopher of the animal rights movement, and how he put into currency an important and evocative phrase: that animals such as these were the "subjects of a life." In other words, they had a biography, a history. I saw the lives of these chickens, and "life" was not the right word to use for what they had. It is my self-appointed task to think about what animals are feeling. I think about it all the time. What were these chickens feeling? Well, panic for one thing, even Tony would admit that. Despair? A sense of futility? Hope that things would get better? I know what I felt: I agreed with my son Ilan, who was accompanying me: "Please let me out, I feel sick."

Starting up a shady hillside in rural Sussex this spring, I startled a golden pheasant, a bird closely related to the wild chicken. I was near to him when suddenly the clouds parted and a shaft of sunlight caught his neck. I was practically blinded by the sheen of dark-blued gold that reflected off his gleaming feathers. He watched me, and I had the distinct impression he was attempting to size me up, to determine whether or not I was dangerous. I saw the urgency of his research. Hidden in the bushes I could glimpse the dark liquid eyes of at least five chicks watching their father watch me. I was an intruder, and we both knew it. I had the good sense to back away as quietly as I could. I saw the bird visibly relax. I felt privileged merely to have seen a wild fowl living the life it was meant to live, with no nightmare visions of battery cages and certain death on an assembly line.

In the wild, both hen and cock elude their enemies, form intense friendships, protect their brood, and greet the golden dawn with a burst of song. This is how chickens and roosters were meant to live.[85]

We, I believe, were meant to protect this life, and to take delight in knowing about it and from time to time catching a glimpse of the joy of pure wildness. It seems a particular cruelty to take a bird designed by nature to live in the wild in a state of wary neighborliness with other birds and confine her to a small cage, deprive her of her ability to give birth in the manner she evolved to do, destroy all family bonds, shorten her life by more than half, and then complain about people who want to restore the chicken to its natural place in the world, not below us, but beside us. A chicken or a rooster can be a friend, who mysteriously permits us to share his or her life for the one small moment we are both on this planet. We owe the chicken the deepest of apologies.

THREE

The Sheep from the Goats

When we think about sheep, if we ever do, the adjectives that come to mind have to do with a certain placidity and lack of agility. In fact, wild sheep are "possessed of tremendous speed and leaping ability, [and unlike goats] they don't have to rely on scaling rock walls to escape a predator; they can outrun it in elegant bounds."[86] Domestic sheep may have lost the ability to run that fast, but some ancestral memory obviously survives: sheep always run uphill if they sense danger. This is doubtless because in their past incarnation as wild animals, that was their only way to escape a predator. Atavistic memories linger: While lambs may occasionally be born in the day, most tend to be born during the night even in modern domestic breeds; this allows them to gain strength before daylight, which would have been of survival value in the wild.

A sheep will respond to his name being called much as does a dog. Elizabeth Arthursson, who has lived for many years with sheep and loves them beyond all measure, simply because they are sheep, remarks on how comical they look when she calls them by name. They race toward her, jumping through the clover with "all four feet a few inches off the ground at once."[87] As for playfulness, few other animals come to mind as quickly as lambs, gamboling about in a meadow or playing king of the castle. The babies leap up, do little dances, and group chase. Like dogs with their play bow, lambs even have a specific gesture to invite another lamb to play: they leap vertically into the air kicking out with their hind legs, an unmistakable plea: "Come play with me."

A common misperception is that sheep lack intelligence. Here is a typical comment from an older book about domestic animals by a dean at Harvard University: "Sheep and goats, like other herbivorous species, belong to the great class of dull-witted mammals in which the intellectual processes appear to be almost altogether limited to ancient and simple emotions, such as are inspired by fear or hunger. They are characterized by little individuality of mind."[88] Fortunately, the pointless human activity of constantly measuring other people's or other animals' intelligence is beginning to give way to the more sensible view that every animal is as intelligent as it needs to be to survive in its world. This is a point made with great forcefulness by the psychologist Euan Macphail, and often repeated by the writer Stephen Budiansky.[89] Keith Kendrick from the Babraham Institute in Cambridge, England (a charitable institution devoted to biomedical research), has shown that sheep process facial images in similar ways to us, and therefore "they are capable of some level of consciousness."[90]

I asked Dr. Kendrick for his opinions on sheep's visual discrimination. He told me, "Sheep make similar use of complex visual cues from the face to recognize each other, and other familiar species, and have very similar specialized organization within the temporal lobe of the brain to assist in this important social recognition task. Overall we have estimated that they can recognize at least fifty different individuals, although in reality the actual figure is probably much higher than this. They can also remember associations with specific faces for several years. Thus their recognition and memory abilities using faces are remarkably similar to those of humans." We should have suspected that sheep have excellent vision, if only because they eyeball us at considerable distance (as a potential predator of course), and because young lambs merely a day old recognize their mothers and reject strange ewes. Since this recognition happens at a distance, we can assume vision is involved.[91]

Many people have reported that sheep have good powers of concentration and can watch television. It strains credulity to believe they are watching without any mental activity; after all, paying attention to what they see is already a form of mental activity. An argument can be made that they only do this to feel secure; they need to make certain that no predator is watching them. But why would they remember the faces of individuals in their own flock for years, even when they have not seen them in the intervening time? Their memory for faces must have evolved for a reason. To retain these images in their minds for years involves complex neural activities and argues for a complex brain.

When I visited Professor John Webster, dean of the faculty at the Department of Veterinary Medicine at the University of Bristol in England, we went for a walk, and we passed a herd of sheep, who seemed peacefully grazing, until we came along. The moment we were within about fifty yards, all the heads of the entire herd of sheep went up at the same time. "What are they looking at?" I wondered aloud. "At you and me," explained Professor Webster, "the ultimate predators. They, after all, are the ultimate prey. If we move back a few feet, they will take up grazing again, but if we go even closer, they will move off." Clearly, we were their enemies. Konrad Lorenz would have said this was innate and, in a wild species, this would be true. But was the ancient fear of man so deeply ingrained in these domesticated animals that it was now genetic, impossible to overcome even with experience? Most scientists, I think, would answer this question with a qualified yes.

I went to meet David Fagan, who currently holds the world championship for sheep shearing (something like 800 sheep in 9 hours). After watching him work, I asked him what the sheep thought of being sheared.

"Sheep don't think," he told me.

"Oh surely," I objected, "they must."

"No, not at all," he insisted.

"Don't you think they like some things and don't like others? Feelings?"

He shrugged. Not from his point of view. But from what I could see, the sheep did not like being sheared one bit. They are grabbed, held tightly between a person's legs, shoved down a chute once they are sheared, and in general treated pretty roughly. I guess if you do what David Fagan does all day every day for years at a time, you forget that you are dealing with a living being.

At New Zealand's Sheep World, just north of Auckland, where people learn about sheep, I met David Oldfield, who has a farm with two thousand sheep on it. He was a young-looking fifty-year-old; brawny, funny, with a wicked twinkle in his eye. We were not on the same wavelength, it was immediately apparent, but we enjoyed each other's company. He explained to me that he sheared twice a year, once in the spring, and once just before winter.

"But won't they be cold and uncomfortable?" I asked.

He had a two-word answer, and it came up a lot in our subsequent conversation: "Who cares?" He was not joking. "They are just sheep," he kept telling me—this from a man who liked them more than most! Later he explained to me how he castrated the sheep and docked their tails—all without anesthesia.

"Doesn't it hurt?" I asked him.

"Sure."

"Don't you care?"

"No."

Some of the bluff exterior was not entirely without redeeming features, and not entirely real. He had young lambs that he bottle-fed. How did he feel about killing them? "Oh, these are not for killing," he conceded, with what looked like—well, sorry—a sheepish smile. "My wife won't let me. We have now about thirty, and they just hang around the house, like dogs. Funny thing is, they seem to like to be with other hand-raised lambs rather than the ones from the flock who are not hand-raised. They seem to find something in common with them. Puzzles me." I left hoping that I had left him with just a bit of curiosity. I promised to send him a copy of my book.

Perhaps New Zealanders have such a difficult time admitting they feel any sympathy of compassion for sheep because of the significant role sheep play in the economic life of the country. I was told by some New Zealand sheep farmers that sometimes a particularly smart lamb will learn to undo the latch of a gate, evidently not an uncommon skill, and the sheep farmer then worries that the lamb might teach his less clever companions to do the same. These particular sheep farmers adduced this as evidence for the intelligence of sheep, which is not something all of their peers would admit to, but unfortunately they were about to reveal a strange (and to me, repulsive) practice.

"What do you do with sheep who can undo a latch?" I innocently asked.

"We shoot them, so they can't pass on their knowledge." Others nodded in agreement. They all had anecdotes about particularly intelligent sheep who were shot as a reward for their cleverness.[92]

We have already seen how people react with compassion to news that a particular animal destined for slaughter manages to escape. I cited examples of cows and pigs in this situation. But I am afraid that the same sympathy is not shown in New Zealand for the occasional sheep who manages to avoid her usual destiny on the way to the slaughterhouse. In April 2003, the *New Zealand Herald* reported just such an incident. The sheep was taken to slaughter, but because there were problems in the plant, it was put out to graze in a nearby pasture. When a Wellington City Council contractor turned up to collect the sheep, it spooked, punched a hole into the clear plastic roof of a nearby greenhouse, and careered through two neighboring properties before being caught. The residents insisted on roasting the lamb in compensation for the damage he caused. It never occurred to them to spare his life.

Fortunately, this attitude is not universal. Carole Webb, who founded the Farm Animal Rescue sanctuary, on the outskirts of Cambridge, England, primarily for sheep, was a sheep lover of a very serious kind. I don't think I have met anyone to whom sheep were dearer. One of her favorite sayings is by Gandhi: "To my

mind, the life of a lamb is no less precious than that of a human being." Considering what Webb has been through herself, this was little less than a miracle: Her husband was killed by a drunken driver;
her daughter died of a heart attack at thirty-two; and six weeks
later, her mother died. Rather than send her into a terminal depression, these tragedies made her more acutely aware of the suffering of others, including disabled sheep. Many of the sheep at her
sanctuary had various disabilities, such as Remus, a three-legged
ram who seems unaware of his disability. Luckdragon, a sweet-
tempered little ewe was born prematurely and lay unnoticed in a
hedgerow when deer-hunters startled her mother. The mother did
not survive, but Luckdragon is now thriving at the sanctuary. Webb
knew the name of every sheep she was caring for (at least a hundred), and what I was able to see with my own eyes was that the
sheep felt similarly about her. Now it is true that it is very difficult
to read love in a sheep, at least for me and most people like me, that
is, people who have had little experience of intimate contact with
sheep. But while I could not determine if what these sheep felt was
similar to human love, I was able to see that they were not only not
afraid of people, but actually sought them out. I was able to walk
up to these sheep and caress them, stroke them, sit down among
them without causing the slightest degree of panic, or even, seemingly, annoyance. Was it just that the sheep were used to people?
Rather, I believe that they were making judgments about the degree of trust they should confer on particular people. Of course
they were using their own experience, which told them that they
had a history of observing people behaving like . . . well, more like
sheep than people. We were suddenly trustworthy. We had proven
ourselves.

How do I know that? How can I, or anyone, know what an animal is really feeling? Of course we are guessing at the specifics of the feelings, as to precisely how similar they are to our more familiar human feelings. It is not a guess, however, that animals have feelings. The distinguished neuroscientist Jaak Panksepp from Bowling Green State University writes that "There is overwhelming evidence that other mammals have many of the same basic emotional circuits that we do . . . Indeed, the evidence is now inescapable: At the basic emotional level, all mammals are remarkably similar."[93] We wonder about our abilities to read the emotions of other animals, but can other animals read human emotions? The answer is easy: Of course they can. Not always correctly, not always exactly, but they can do it. They assess our moods, in fact we can claim that they spend a great deal of time, dogs do, reading our moods, trying to discover what is likely to happen to them. Things would go badly for them if they are wrong most of the time. The truth is that they are right, most of the time. Sheep, too, like any domesticated animal, must engage in this exercise much of their lives. To do otherwise is to risk constant danger.

I don't think my experience in Cambridgeshire was unusual. Shepherds who live with their herds for long periods of time can confirm what I saw. Unfortunately, shepherds rarely are asked, or volunteer information about their feelings for the sheep they live among and even less about the feelings of the sheep toward them. When I e-mailed a retired shepherd to ask him about the feelings of sheep, he cut off any further contact with me: "I don't humanize animals." Fortunately, though, not all shepherds feel this way. The shepherds who live with their sheep, in the Mediterranean tradition of the *transhumance* where sheep are taken into the high mountains during the summer and return to the valleys in the fall, are something of an exception. Many of these shepherds have been doing this in a family tradition that reaches back many centuries, and have developed strong feelings for their sheep. One sees this still in the south of France, in la Camargue, for example.[94] One of these shepherds said: "Sheep are creatures who obey dogs, they obey shepherds and they obey the stars."[95] When an old shepherd was asked his age, he said he did not know exactly. Asked how many sheep he had, he easily answered 223. "How is it that you have retained the exact number of your sheep but have lost your memory for the number of years you have lived?" "Well, whatever happens, when I return to the farm, my years will still be with me. Such is not the case for my rams and ewes."

As the most populous and widely distributed domestic ruminants, sheep (*Ovis*, the domestic variety is known as *Ovis aries*) are one of the most heavily exploited of all domestic animals. They extend from the Arctic Circle to the most southerly tip of South America. Sheep and goats were first domesticated in western Asia during the eighth and seventh millennia B.C. From the beginning, and no matter where, they were and are exploited mercilessly. In Russia and in some eastern European and Middle Eastern countries, sheep produce milk[96] as well as wool, and are slaughtered for their skins and for meat. In Tibet and other parts of the Himalaya Mountains, sheep produce all these commodities and also serve as pack animals. (In addition, their dung is used for fuel, their guts as sewing thread, and their horns as needles and as trumpets.)

A young lamb will imprint on a human, if he encounters some-one hours after he is born. A woman coming across a young lamb in ancient times might well have nursed the lamb. In this way, domestication may owe its existence to women. At first, sheep were not domesticated for their wool, since in the wild, they have a hairy outer coat which has to be made finer by breeding before a fleece suitable for textile use is obtained. The idea of using their wool probably only came later, after it was observed how the wool would come off the animals in large wads. We don't know if sheep were used first for meat or even for milking (sheep cheese is still common).

Sheep are said to be animals "predestined" for domestication. All mountain breeds tend to stay in one place, for example, a trait that is called hefting. This instinct acts a bit like imprinting, except that instead of being imprinted on a person, a sheep becomes attached to the place it was born. A mountain sheep will, I understand, graze within a few hundred yards of its birthplace all its life. This is the reason mountain farms are sold with their native flock. It also allows the shepherd to find where any particular sheep is likely to be graz-ing.[97] The ease with which sheep can be handled, and their obvious vulnerability, has produced not compassion on the part of humans, but disdain and an eagerness to exploit.

Dogs were certainly domesticated before sheep, and it may fairly be said that "the first control of wild sheep may have been affected with the aid of the dog well back in the pre-agricultural stage of eco-nomic evolution."[98] Brian Vesey-Fitzgerald, in his book on the his-tory of the dog, points out that the greatest achievement in canine history was the creation of the sheepdog: "In training a dog to hunt or to guard his property, all that man had to do was to encourage the dog's natural instincts. But to train a dog to herd and care for sheep or cattle meant achieving a complete reversal of the dog's natural instincts, for the dog is a carnivorous animal and a hunter, and these are his natural prey."[99]

Sheep have always been more responsive to other animals than to humans, sheep-herding dogs being an obvious example. They mostly achieve their results through fear. The sheep respond to the dog, either to his barking or to his eye, because they look upon dogs as their natural predators (which, in fact, they are). A good dog, of course, never hurts the sheep; his aggression is under control, but it is nonetheless real aggression. The sheep-guarding dogs, on the other hand, go about their work because they love the sheep, or at least they have affection for them. Whether the dogs think they are sheep (which is what those who train them believe, since they raise them as puppies in the presence of sheep), or whether they only behave this way out of a desire to please their true love (humans) cannot be definitively answered—though I suspect the latter is the case. Sheep also respond to other sheep who have been especially trained by the shepherd to lead the flock. There is a practice in contemporary slaughterhouses of using a tame sheep to entice the others to come in. In an instance of human candor, this animal is called the Judas lamb in recognition of the treachery that she engages in, though it is not likely that she is aware of her true function. A seventeenth-century medallion by Julius Wilhelm Zincgreff shows a flock of sheep leaping from a bridge to cross a river, led by a tame ram with a bell around his neck. Janet White, who had a passion for sheep ever since she was a child, eventually became a shepherd. She found it easy to send sheep to slaughter but noted a tendency that almost all shepherds have of keeping one particular sheep safe: "Our current mascot is a placid bottle-reared wether [castrated male sheep] who appears to regard humans as friendly two-footed sheep. It would be too treacherous to send him to market. He would walk up to his own executioner to say hallo."[100] Italian sheepherders in the Abruzzo in central Italy used what they called a *guidarello*, a castrated ram, to herd the sheep. He would respond to commands of the shepherd, always had a name, and slept with the shepherd from the time he was small. The shepherd would use the body as his pillow, thereby increasing the bond between them.[101]

But sheep form bonds with other animals as well. When I was visiting the Midwest, I went to Ashgrove, in rural Missouri, along Route 160 north of Springfield, to meet Mary Hurt, who lives on an eighty-acre farm. She had a cow, named Whisper because when she called she sounded as if she were whispering. Whisper was born blind. When the neighbor in whose field she was born wanted simply to leave her to die by herself in the field, Corey Hurt, Mary's twenty-year-old son, carried her for half a mile in the rain back to his house. To everyone's astonishment, Rammo, a macho two-year-old Rambouillet ram, took to her. Rams tend to be loners, and he was a pretty tough ram, so it seemed most unusual that he would take up with a blind member of another species. But he did. Mary told me "somehow he felt he needed to protect her." And protect her he did. He would graze next to her all day and guide her about the field, making certain she did not bump into the fence or posts. She was an extremely affectionate cow and loved to stand by when the children played volleyball, often attempting to join the game. Corey told me he could remember these games very well because country music would be blaring out of the radio, which for some odd reason seemed to attract the chickens who would come to listen. The geese were in the pond, with their leader who had a bad foot and a broken leg, but was amazingly spunky. When Whisper's name was called, she would look around expectantly. She did not seem to be aware that she was in any way different from the other animals. When she had a calf, Shout, sired by an Angus bull, Rammo behaved paternally toward the young animal, more so even than to his own offspring, several bouncy lambs. Whisper lived to be four years old and then died in 1996 of a viral infection. Rammo mourned her a long time, standing by her dead body, calling and calling. This reminds me of the words of the classical author Aelian: "It fills me with shame, you human beings, to think of the friendly relations that subsist between animals, not only those that feed together nor even those of the same species, but even between those that have no connection through a common origin."[102]

Is there something about the innocence of sheep that stimulates men to ever-greater heights of cruelty? The apotheosis seems to have been reached in Buenos Aires in the nineteenth century: "Wool was formerly so scarce at Buenos Ayres, and cattle so plentiful, that sheep were actually driven into the furnaces of lime-kilns, in order to answer the purposes of fuel. This fact could hardly have been mentioned as credible, however undoubted, if a decree of the King of Spain, prohibiting this barbarous custom, were not still preserved in the archives of Buenos Ayres."[103] Some people, on the other hand, were attracted to sheep for these same trusting qualities. Elizabeth Arthursson, whose book I mentioned earlier, wrote, "There was something infinitely pleasing to me about keeping an animal that lived off flowers and not the dead bodies of other animals. I had always liked a sheep, but until I had actually owned these two had never imagined just how delightful they were. They were warm and gentle and loving, and there was nothing about them that one could possibly dislike. Most things seem to have disadvantages, but sheep to me were perfect."

Germaine Greer, who keeps five North Country Cheviots, likes and respects sheep. She disagrees with the usual judgment that sheep are simply "lumpish animals that have no wit to comprehend their own misery." She writes that "Sheep, it seems, are ours to grow, like any other crop; emblems of animal passivity, to be sacrificed by divine edict, their blood dashed upon our doorposts. It made sense that the first mammal to be cloned would be a poor, bloody sheep."[104] Few people have attempted to live with a sheep as a member of the family. But those who have, the biologist Charles Hansen, for example, claim that they are easy to train and learn quickly, "perhaps more quickly than most dogs."[105] He said that when they tag along, they "often reminded us of our own children."

Little study has been done on the sounds made by sheep; we tend to think of them as monotonous. Clearly, we are the ones who are tone deaf. Those who are more in tune hear more. Two writers vividly describe what they heard on a moonless night in March in the 1960s: "We came directly upon a large band of ewes and lambs bedded among the Joshua trees on a knoll at the edge of the wash. We shut off lights and motor and listened for 15 minutes to a veritable din of anxious bleating, calling, and answering, which gradually subsided into quieter and plainly recognizable tones of recognition and reassurance, then finally silence when the last anxious mother touched her lamb in the dark."[106]

The argument has been made that sheep, in particular, need humans to survive. Wild sheep may seem happier and healthier than our domestic sheep, but this, to at least one scientist, is an illusion. Wild sheep on their own are subject to horrible diseases and need our intervention to thrive, or so claims Colin Tudge. In an article about how farm animals depend upon humans for their own welfare, he writes, "Lakeland hill sheep may appear to enjoy an idyllically Rousseauesque freedom. In reality, they may be besieged, indeed eaten alive, by blowfly maggots in summer, and die of cold or starvation in winter."[107] I find it hard to believe that any wild animal normally dies of cold or starvation in winter. It would be a strange trick of evolution to produce an animal who could not survive the conditions into which it is born. Surely such a species would disappear in a few short generations. Of course if these animals were feral and were not really meant for the conditions under which they have been raised, then the argument is valid. And it is true that some domestic animals lose the tendency to seek shelter in rain and hail. E. S. E. Hafez, an authority in this matter, wrote about large flocks of domestic sheep who, in the absence of shelter, "often mass together and smother in severe winter storms." But this does not happen to wild mountain sheep because they live in rocky environments "where natural windbreaks and shelter are provided."[108] In other words, nature has provided what we removed. To argue that animals are better off in the prison of our making, even if that prison were to be a benign one, is curious.[109]

But Stephen Budiansky, in his book *The Covenant of the Wild: Why Animals Chose Domestication*, makes just such an argument: "One moral lesson that the natural historical view of domesticated animals can teach us is that just as sheep may 'expect' to be killed in their symbiotic relationship with humans, they also, by virtue of their evolutionary 'choice' thousands of years ago, 'expect' to be better off than in the wild."[110] I am aware that Budiansky is using words like "choice" and "expect" in a purely metaphorical sense, or merely as shorthand, but the idea seems fundamentally flawed. No mammal, human or otherwise, would wish to live a trouble-free life if the end result is an early slaughter. I have never met a human who would.

Domestication has led to many physical changes in sheep. Compared to their wild forebears, domesticated sheep have smaller hearts and smaller eye sockets, and their brains are 20–25 percent smaller, too.[111] The bodies and horns of domesticated sheep are smaller. The tendency to moult has disappeared, and now most are white. Wild sheep are white only on their bellies and a dull brown color everywhere else. (There is still much that we don't understand about sheep, such as the strange—and still unexplained—fact that white sheep, like white pigs, suffer from, or even die from eating buckwheat, whereas black or dark-wooled individuals are not in the least affected, a fact that long ago intrigued Charles Darwin.[112]) It is possible that they have lost some of their immunity to lungworms, tapeworms, and intestinal parasites, the bane of sheep farmers today. This means that the major diseases tormenting sheep today have been introduced by domestication; in their natural state, in the wild, sheep would be at least partly immune and therefore lead healthier lives *without* medical intervention or drugs.

It has been argued by some animal scientists that sheep are more stoical than humans, not just in their appearance, but in reality, that they do not feel pain the way we feel pain. This argument goes back to ancient times and has been invoked by scientists and farmers alike. Is it true? I asked John Webster, perhaps the world's leading authority on the physiology of pain in animals. He produced an article he had written about this very matter, in which he writes, without fudging the matter in the least: "There is clear physiological evidence that the intensity of pain sensation in cows and sheep is similar to that in man." Perhaps, though, I suggested, sheep and other animals are able to adapt to pain much better than we do, and after a while simply cease to feel chronic pain, for example. Quite the contrary, according to Professor Webster: "Chronic pain causes changes in the processing of pain signals within the central nervous system which amplify the sensation from the damaged site. Far from adapting, the sensation gets worse with time."[113]

It is true that sheep are subject to any number of diseases and parasites: foot rot, cutaneous myiasis, and sheep scab. They are tortured by sheep scab; the mite gets under their skin and causes them intense irritation. Being dipped once a year is the antidote. Of course they are dipped not to save them the irritation, but because the mites cause such anguish that the sheep rub off their wool and sometimes their own skin. I do not know, and have not been able to find out, how prevalent sheep scab is in wild sheep. Most of the diseases domestic sheep are prone to—which cause them such agony—do not afflict wild sheep, who have evolved to withstand normal conditions. Some shepherds make the strange but compelling claim that sheep have a death wish, their one ambition being to die before their time. They get foot rot[114] and blowfly (a horrible, painful disease in which sores are infested with maggots), and tangle themselves up in the brambles. I call this view strange and compelling because it raises the possibility—a dim one, granted—that sheep long for their previous state where they were healthier, wilder, and more free. This is a disturbing thought: that one of our domesticated species prefers to die than live in slavery. The life we provide sheep is not a natural one, and we must recognize that the unremitting agony they are subjected to is one of human making. The fact that their lives appear benign to us is a comment, not on the real lives of sheep, but on our own blindness.

Why is it that we get sheep so wrong so often? Partly it is that they don't look like us. A pig has eyes like ours and makes sounds that we recognize as "human." Sheep have facial expressions that we find hard to read and seem to bear pain with so much fortitude that we think they are not like us. Somewhere, of course, we must understand how deeply vulnerable they are, or we would not have the expression *Agnus dei*, "the Lamb of God," to describe Jesus Christ: "Behold the Lamb of God, which taketh away the sins of the world." (John 1:29). Surely one of the reasons we use this expression is that we recognize the innocence of the animal, how a lamb poses no threat to anyone. Yet it is killed. The expression "lamb of God" also refers, clearly, to the great love an ewe has for her lamb, something that goes unrecognized except in this expression. The Bible is littered with references to sheep as both animal and symbol—think of the Good Shepherd to describe God—and as sacrificial beast.

Matthew Scully pointed out to me a beautiful passage in the Old Testament (II Samuel 12:3) that makes it clear that we are not the only ones to feel deep attachment to animals we otherwise eat:

"The poor man had nothing, save one little ewe lamb, which he had bought and nourished up: and it grew up together with him, and with his children; it did eat of his own meat, and drank of his own cup, and lay in his bosom, and was unto him as a daughter."

Children who grow up on farms have strong feelings about the animals they live with. Susan Street, a twelve-year-old girl from New Zealand, gave a lyrical if not entirely grammatical description of how a child feels in this passage about a goat named Tinker Bell: "She lies on the hay and I lie on her. The rain drips off the roof while the rain pours outside; she's warm and I like just lying there talking to her softly. It's only a small shed about four feet high and we're both squashed together but it's nice just how I like to feel."[115] There is a special relationship between lambs and children, expressed in the nursery rhyme about Mary having a little lamb. It is not surprising, therefore, that there has also been an awareness that not all is well with this relationship. (Is it at all possible that the expression "sheepish" to express shame derives from this? There seems no other obvious reason for the expression.) Children who become close to lambs often reflect with a feeling of sickness that this relationship is then destroyed by adults in a particularly treacherous way: by killing their companion. Iona and Peter Opie, in their 1959 book *The Lore and Language of Schoolchildren*, quote a frightening version, much closer to reality, of the nursery rhyme that reads:

Mary had a little lamb
Her father shot it dead
And now it goes to school with her
Between two chunks of bread.

We think of lambs as innocent, but we think of sheep as dumb, passive, and ignorant. How stupid can a sheep be, though, when "sheep and goats can distinguish between different strains of the same plant species so similar that a botanist might find the same task difficult?"[116] Some scientists believe that sheep show what has been called "nutritional wisdom," that is, given a free choice of food deficient in certain minerals, the sheep select and balance their diet, correcting for the deficiency. The wild sheep was known by the earliest authorities to be "an intelligent and courageous animal, capable of escaping, by artifice and swiftness of flight, from his larger foes, or of beating off his smaller enemies by a dexterous use of the weapons that nature has given him." This was the opinion of William Youatt, whose authority has been invoked on a number of occasions in this book. He went on to say that this might not be so obvious in the domestic sheep because "our connection with sheep extends no further than driving him to and from his pasture, and that at the expense of much fright and occasional injury, and subjecting him to painful restraint and sad fright when we are depriving him of his fleece" all of which leads to the depression of the intellectual powers of sheep.[117]

It will come as a surprise to my readers, as it did to me, that leadership in a flock of sheep does not go automatically to the strong, large male, but on the contrary, belongs to the older, smaller, even frail female. This was a point made long ago by the great animal behaviorist, J. P. Scott: "Leadership of the flock went to an elderly ewe, inferior in strength and fighting ability to almost any ram, and often inferior to the younger ewes. The position was achieved mainly by the care and feeding of her descendants without, as far as the observer can tell, any instance of violence toward her offspring."[118] Not a bad model for humans as well as sheep!

How different are sheep from goats? They are different species, different animals, but many people still have trouble telling them apart. People know that sheep produce wool, but few are aware that mohair and cashmere come from goats. It is easier to tell apart the males: the male goat has a beard and a caudal scent gland that makes him disagreeably smelly to humans, though not of course to she-goats. A billy's smell is intensified by his habit of bending his head between his forelegs to catch his urinary spray at the time of rutting, for two months of the year, usually November and December, when hormones in the bloodstream activate the scent. All goats have erect tails, whereas those of sheep are pendent. The horns of a goat are more slender than those of sheep and rise vertically from the top of the head (compared with the V-orientation in sheep). In India today, it is still the fashion to breed sheep and goats who look alike, so that only on close inspection can they be distinguished.

In their behavior, goats and sheep are often remarkably similar, especially among their young. Play is important in the life of all young goats. Kids[119] (the young of goats—the origin of our word for human children; again, probably because of their playfulness—related to "no kidding"!) gallop in circles; make high, arching jumps and little dances; and engage in social play with the entire herd taking part. They jump, they slide, they leap onto their mothers' backs, toss their heads, whirl on their own axis, spring vertically, and play fight one another for hours. I already noted that goats appear independent to us; whereas sheep definitely have a tendency to follow a leader—a "bellwether" with a bell slung around his neck, usually a black animal, as the leader. I don't know if the color black makes these bellwethers more distinctive to the sheep, or if they are partial to the color or perhaps even in some awe of it. Goats seem naturally to be used as bellwethers, and sheep will follow them. Professor Webster watched a billy goat on a leash play fight with a German shepherd for an hour. It was pure play. When I talked with him about sheep and goats, he showed great enthusiasm as he spoke of the extraordinary personalities of goats, their dignity and lack of fear of humans. A goat will butt you for a joke. They have a sense of humor, as do dogs and cats, but which seems lacking in cows and sheep. While cows are cautious of human contact not familiar to them, goats tend to seek us out, and for some mysterious reason do not perceive humans as a threat.

The domestic goat is directly descended from the wild goat, or Bezoar, distinguished for its scimitar-shaped horns. We call the she-goat a doe or nanny goat, and the he-goat billy or buck. The goat was the second animal to be domesticated, after the dog. Between 7000 and 8000 B.C., people in the Middle East took the Bezoar into captivity, choosing it over other animals because of its eating habits, which are still notorious today as cartoons of goats eating the laundry show. Given a choice, goats, like every other animal, choose the food that is best suited to them, but they seem far less picky than many other animals. Actually, they are picky. It's just that they can, as others cannot, digest high fiber and lignon-content diets. Since herds of goats would eat everything in the vicinity of their nomadic masters, the tribes would move on, without bothering to replant any of the trees or other vegetation destroyed by their goats, and in this way it is believed that the goat created the deserts of the Middle East and Sahara. I'm reminded of a photograph in *A History of Domesticated Animals* by F. E. Zeuner of a large number of goats in the Moroccan desert who, having climbed a tree, are in the process of eating it into annihilation.[120] Certainly goats are known for their ability to live in arid lands, which support only xerophytic (such as cactus) vegetation that would defeat just about any other mammal. There are about 700 million goats in the world. They thrive in almost as wide a range of temperatures as humans, and more reasons for their ancient popularity are the facts that they are easily managed (even by children), their milk is easy to digest (especially for children), and they provide meat and fiber for clothing and housings, as well as skins for clothing and lightweight, watertight containers, so important in arid desertlike conditions.

There are many negative associations with goats. Even the common term "scapegoat" refers to a bizarre ritual in which "two goats were brought to the altar of the Tabernacle and the high priest cast lots, one for the Lord, and the other for the powerful, high-ranking evil angel Azazel. The Lord's goat was sacrificed, the other was the scapegoat; and the high priest having, by confession, transferred his own sins and the sins of the people to it, it was taken to the wilderness and suffered to escape."[121] Who can forget Goya's *Sabbath, or the Gathering of Sorcerers*, where the he-goat monster dominates the center of the canvas, and an old witch is offering him the sacrifice of a child? Indeed, at the Last Judgment, goats are on the left, the sheep are for salvation, for they have heard and answered to the voice of the shepherd (the origin, probably, of the expression "sorting the sheep from the goats"). What this acknowledges is the amazing independence of goats, who are unpredictable, and in human judgment, self-centered. (We think they are independent when we compare them to sheep, because they will not invariably follow a leader, and we think of them of them as selfish because they are not always ready to obey us and find many ways of avoiding doing what we want them to do. Strange that we think raising these animals for our own use does not demonstrate selfishness on the part of humans!) Goats are flamboyant and extroverted, and have playful minds, many people who live with them tell us.[122] For many humans, goats remind them of other humans: curious, greedy, and proud. Perhaps it is because we recognize so many human features in goats that, by and large, our myths about goats (with the exception, a major one, of the relation to the devil) are less negative and aggressive than for just about any other domesticated farm animal. We think of them as canny, which indeed they are. Some of our prejudices are not even far wrong. For no animal is as sexually precocious as the goat. Kids have been mated at three months by billies of under a year old and produced young. One male can easily mate with ten to fifteen females in a single day.

Maybe the reason goats do not fear humans is that goats fear few animals. In India, goats have been seen to kill leopards. In many ways, they remind people of cats. They can revert to the wild and quickly go feral. More so than sheep or cows, we treat goats more as a pet than a farm animal whose only value is for exploitation. There are numerous breeds of goats raised for pure pleasure, valued for their aesthetic look rather than their meat or milk. From an early time, too, some goats developed lop ears and even came in dwarf versions. One breed of dwarfs from Khartoum in the Sudan dates from 3300 B.C.

There are two types of fleece-bearing goats: Angora or mohair, native to the province of Ankara in Turkey (hence its name), and the

Pashmina (from the Persian for "woolen") or Cashmere (the old spelling of Kashmir, where the Europeans first encountered the fiber) found in all central Asian mountainous regions, including the Tibetan plateau, Kashmir, Mongolia, and China. Astonishingly, only 100 to 200 grams of wool can be combed out of each individual goat, which is why this goat fleece has been precious from antiquity to the present day. Since they live above 15,000 feet, "for warmth these white, long-haired goats, with their high twisted horns, grow a double-layered coat consisting of a fluffy lining of wool covered by a waterproof shield of hair. The finest wool, softer than any sheep's, comes from goats living in the highest mountains. Traditionally the wool was combed out in spring over a period of 8–10 days and primarily woven into soft shawls bordered by beautiful vegetable-colored, curving patterns which became fashionable in England" and remain so today.[123] Because the Angora goat has a "baa" like a lamb, early writers supposed it was a cross between a sheep and a goat, which it is not.

I went to visit a retired biologist, Dave Needham, and his wife, Margaret, who live on a ten-acre property in the Kaipara Valley, just north of Auckland. We took a walk with Pedro the dog, well-known for his kindness to all living creatures, Tarquin the Siamese cat, who delights in playing with the goats, and Solomon and Sheba, twin dwarf Nubian/Angora-cross goats. The goats were as eager to follow us as the dog and cat were. They went wherever we went and could hardly wait to investigate wherever we stopped. We walked to an old chicken house and, within seconds, both goats were inside, sniffing, looking, touching. They kept coming over to us to have their necks scratched. Dave told me that what he loved most about them was that they were full of personality. Being around them, walking with them, and taking hikes with them was never boring. In the short time we were walking along the stream with the goats, I could see how you could easily begin to see the world from the perspective of a goat. The only thing I missed was some large rocks for them to climb. Dave was aware of this deficit and had built a wooden play structure that they could climb. We walked over to it and they were like kids (which of course, they were), eager to show us their home playground. It was a beautiful sunny day, and the fields were shining emerald green. Had we not been there to distract them, the goats would have been lying soaking up the sun and waiting for cooler weather before they began their daily play. Dave got as much pleasure from his two wonderful goats as he did from his dog and cat. They clearly reveled in the attention.

Do goats eat tin cans? Of course not. They enjoy the bark from trees and most pulp products, so they chew the paper from the cans. Wild goats are very particular in what they eat, since so many plants are poisonous or harmful to them (such as rhododendron, azalea, mountain laurel, buttercup, cowslip, even lily of the valley). They love thimbleberry and salmonberry, in addition to blackberry leaves, blossoms, and berries. If you offer your goat spoiled food, he would rather starve to death than eat it.

Recently I went to see Professor Marilyn Waring at Massey University in Auckland, where she teaches public policy at the School of Social and Cultural Studies. Her name is familiar to everyone in New Zealand, for in 1984 as a member of Parliament, she brought down the conservative government of Robert Muldoon. She left the country for the election period so that the focus would be on the election and not on her. When she returned, she wanted to put her dream of living a rural life into reality. She wanted to recover from Parliament by living on a farm in the best of company: goats. She bought a farm first in Port Albert, then in Wainui, just outside of Auckland, and two hundred goats.

"Why goats?" I asked, though I suspected I knew the answer.

"I had been pushed around for years, and I decided I didn't want it any longer. Goats would accept me as head goat and seek to be close to me whenever they could. Moreover, goats are funny, fun-loving, inventive, fun to be around, cute to look at, they have good memories, like to tease, and oh yes, they are wise. Very wise in fact." She thought for a moment, and then added, "They also give you unconditional love."

"Like a dog?" I prompted.

"They are smarter and more interesting than dogs," Waring said. I found that my dogs always knew when I was trying to help them recover from an injury or an illness, and I thought that was pretty smart, perhaps even wise. Goats are like that too, according to Waring: "When I help a goat, she never forgets it. I had a wonderful doe, Jean was her name. She was very concerned about her friends, and would keep an eye on them. She was an angora with long, beautiful hair known as mohair. Often her friends would get their long fleece tangled in blackberry bushes, and the more they struggled the worse the situation. Jean knew exactly what was wrong, and would run back to the farmhouse and start bleating." Waring could recognize many different bleats, and this was the "please help" bleat. She would follow Jean and find a distressed and helpless goat, patiently awaiting succor. The help was even more deeply appreciated when the does were kidding and would sometimes have trouble. The sound of a baby goat in difficulty is deeply disturbing, because they cry exactly like human babies. Waring would sometimes become a goat mid-wife, and the mothers would lick her hands in what could only be described as appreciation. They were clearly grateful and never forgot the help. "Be polite," they seemed to tell their kids, "she was there for you." Sure enough, though goats love to butt, they never butted Waring.

Waring and I both love the fact that goats have a crèche. Does nurse their young kids only three or four times a day and like to get away from their hungry youngsters. They will suddenly slip away and find a rich pasture. But they are not in fact abandoning their kids. When mothers are busy eating, several young goats are herded together by two or three "aunties" who stand by to make certain that they come to no harm, do not stray too far away, and are never left alone. If things get too difficult for them to handle, if there are any serious problems with the kids, the aunties will bleat out to the other goats, and Waring knows something is wrong: "They bleat in a special way, and believe me, I know what it means. I can tell from the sound whether they are having fun, are getting angry, or are desperate for me to come and help them."

"The playfulness of goats," she tells me with a nostalgic laugh, "must be seen to be believed. I remember once watching them running downhill to a corrugated roof which was followed by a few meters of bare ground, then another roof. How they loved to hear the sounds of their hooves on the roof. The thing is, they would actually take turns. They did not all come at once, just one a time. The others would wait in line, and they reminded me of nothing so much as kids on skateboards, waiting their turn for the fun. I laughed aloud, and I think they knew why I was laughing and that just increased their pleasure!" They play games with Waring, such as when they would push her to chase them around a tree. They peek out to see where she is, then turn the other way. Quick and agile, it was impossible for a human to beat them, which they loved. "I was head goat, you see, and what more delicious pleasure than to beat the head goat?" Before this, I had not realized just how hierarchal goats are, worse than men in the army. Every goat has his or her place, and they will fight to maintain it. Even number 23 wants to remain so, and not slip to number 24.

Games are fun for goats, but being really naughty is best of all. (Is that where we derive the expression "getting your goat"?) Once the entire flock managed to get out of their fenced pasture and into a neighbor's yard, Waring told me. "I was angry, and they knew it. I raised my voice and told them what I thought of what they had done, and said to them in no uncertain terms: 'Now get back here at once.' Well, their ears were moving fast, they got it, they knew I meant business and they ran back through the fence at great speed. They are very responsive to tone of voice. At other times, when I was not really angry, they would play dumb, and pretend they could not possibly imagine how they had gotten out. They could not find the hole, there was no gate as I could plainly see, and so they would love to return home, but did not have a clue how they would manage it." From what I gather, there is no question that goats laugh; they may lack our exact physiognomy for it, but there is no mistaking their intent.

Tofu, Waring's favorite goat, lived to be sixteen, just short of the *Guinness* record for longevity in a goat. Waring was driving down the lane when she spotted that Tofu's mother had not gone inside in the pouring rain. The doe could not get him up on his feet—Tofu was stiff with cold when Waring found him. A twin, his mother did not appear able to nurse him. So Waring bottle-fed him and hand-raised Tofu. He imprinted in a big way. "He was the best company anybody could wish for. He followed me wherever I went. Everything I did was of interest to him. He noticed all that went on. Actually, all goats are good at noticing what is going on around them. He had no fear of any animal on the farm. He was just a likeable goat, and I think all the animals felt about him just as I did. Knowing, as I do, what a loveable animal a goat is, and how deeply sociable, it really makes me unhappy to see so many goats tied up on the roadside, being used as a lawnmower, without a companion. Moreover, some of them don't even have shelter, which is unconscionable. You know, their fleece lacks lanolin, unlike sheep, so the rain simply makes a hole in their fleece, it saturates it, and they get extremely cold. That is why goats hate rain."

I could not help myself from asking, since we were having lunch, and Waring was no vegetarian, whether she ever ate goat.

"No," she explained, "I could not bear it. Cows I eat, and that is why I called the one on my farm destined for the home freezer 'Barbecue,' because I knew her fate.

"But how could I eat my friends? And my friends they certainly were: when I was absent for a time and returned, the entire flock would come running and bleating, frisking about, skipping sideways with joy—there was just no mistaking it: 'We are so glad you are back, Mom!'"

I liked to think that she wouldn't trade a day with her goats for all her time in Parliament!

FOUR

Till the Cows Come Home

We have a strange relationship with cows. We drink their milk. Their skin and flesh are ubiquitous in our lives. We wear leather shoes and carry leather bags. As a nation, Americans consume astonishing amounts of beef. On average, an American eats about seventy pounds of beef per year. In 2001, forty million cattle and calves were slaughtered in the United States for food. Despite all this—or perhaps because of it—as living, breathing animals, cows largely go unseen.

Journalist Peter Lovenheim decided to examine this relationship. The result is a startling book, *Portrait of a Burger as a Young Calf: The Story of One Man, Two Cows, and the Feeding of a Nation*. Lovenheim was in line at McDonald's to buy his young daughter a Happy Meal, which came with an adorable black-and-white Beanie cow named Daisy. It suddenly occurred to him, as it no doubt would to many of us, how can it be that our children are given a cuddly toy cow one moment and then proceed to eat the grilled remains of a real cow the next? This disconnect between what we know and what we eat bothered him to the point that he set out to understand the whole experience by buying twin calves. Calling them number 7 and number 8 to avoid forming the attachment that naming facilitates, he observed the process—repeated on factory farms millions of times a year—of how a calf goes from birth to plate. His is a chilling story, made all the more palpable because Lovenheim was determined that sentiment not cloud his reporting: He intended to watch his own calves be slaughtered.

The account succeeds; his intention does not. At the end, he made a decision—not an intellectual one, he tells us, one from his gut: "I simply didn't want them [his two calves] to disappear and become meat." He would kill them, he said, if he had to, to live or feed his family, but he acknowledges, "Any of us can nourish ourselves, as my sixteen-year-old daughter does, without meat." He has an epiphany at the end of the book, the kind we all hope for: "In one sense, we all have calves in our lives: things that are small, voiceless, or vulnerable, and over which we have power. It might be a child, an older parent, a stream or a stretch of woods, or a Holstein calf." He ends his book with the hunch that once you connect all the dots, "you see that all the calves are number 8."[124]

Certain people have always seen cows in a deeper way than the rest of us. There is Rosamund Young, whose Kite's Nest farm is perhaps the most famous organic farm in England. Kite's Nest lies above the Worcestershire town of Broadway on a tree-wooded slope where the Cotswolds subside into the Vale of Evesham. Rosamund lives there with her brother, Richard Young, and their mother, the formidable Mary Young. They have 390 acres, of which 100 acres is woodland. Rosamund is an amazing person: she has the same relationship with her cows that most people have with their extended family. She knows the names of all 132 of them, Welsh blacks and Lincoln reds, but that is only the beginning. She also knows each cow's complete genealogical history, who is related to whom, and who likes whom as well. Such individuals, that is her main theme. Wendy is terribly friendly, Jenny is the opposite. Lulu is imperious. Janey loves to be touched. Dither is affronted.

I was surprised to see so many cows with horns. Are they not dangerous to one another? Never. They never use their horns for fighting. One particular cow, who was born without horns just has to *look* at another cow, even one formidably armed with a strong set of horns, for her to back down. Jill gathers all her children around her every night. Jane sleeps alone. "Cats and cows are no one's servants," Young told me. Even at birth they are different: some are very mature and behave like little adults, others are babies and remain so for months. One thing that has impressed her enormously is what good practitioners they are of herbal medicine. They will eat the strangest things—thistles and nettles, in particular—when they need to cure, say, an upset stomach. They are quite discerning, willing to walk 400 yards to seek out warm pond water, while at other times they will drink only fresh running stream water.

A cow is so much like a woman, Young told me. They carry their young for nine months, just as we do, and then they suckle them for between nine to twelve months, much like many human mothers. "Do they have a sense of death?" I asked Young. "Oh yes," she told me, "because I have seen that they are worried by the smell of blood. Of course, they have such a developed sense of smell. They recognize one another that way, and probably also learn how another cow is feeling. Every cow will approach a new calf and sniff her, as if getting to know her."

Young thinks that cows can teach us to relax. They have no deadlines. We learn to live more naturally when we are around them. "That is why I like them," she explains. "They are much nicer than we are, more integrated, more whole. They pick up my mood very quickly, and hate it, for example, when I am in a rush. They just don't believe in rushes. I have seen calves playing tag with a fox around a tree."

We walked past the five dusky owls and the bats, past the mouse-eared chickweed and the harebells, looking at the Black Mountains in the distance. We were going to see the early purple orchids. We saw ladies' smock and dropwort, of which only two or three farms in England have patches left. Richard is in love with his farm: "Do you realize," he told me, picking up a handful of dirt, "that in just one teaspoonful of healthy soil there are more living organisms than all the people on the planet?"

It was getting dark. Rosamund said, "See how the cows settle, looking away from the lights of the valley and toward the moon." Rosamond Young definitely finds her greatest happiness among cows. She could sit down and write 132 biographies and "each one would be completely distinct from all the others," she told me. She has absolutely no doubt that cows feel all the major emotions that humans do. "Even worry?" I asked. "Especially worry," she replied. They have so many different kinds of worry, ranging from the most mild, when a calf wanders out of sight, to the most extreme, when they think something terrible has happened to it. "How about surprise," I asked, remembering that some philosophers claim that animals cannot feel surprise because they cannot anticipate the future. "Well," she said, "how about this. My Welsh black cow had six black calves, and then one day she produced a pure white calf (she had been mated with a Charolais bull). She came round to our door and stared at us with a look that was easy to read: 'Are you sure it's mine?'"

Laurie Winn Carlson points out in her recent book about cattle that cows "are nature's most protective mothers" and will attack any animal that threatens their young. (This is true of any number of wild animals as well; we need only think of elephants, for example.) She quotes the essayist Nancy Curtis, who writes about seeing a cow disoriented for a month after losing a calf, "returning to the site of the birth, searching, calling. It tugs at me in a deep spot where the mothering instinct is never completely buried."[125] When a cow loses her calf or is separated from her calf, it is rarely due to any sort of natural calamity. Alas, it is almost always human-engineered. Farmers separate calves from their mothers at birth so that they will not drink the milk meant for them. We want that milk, and farmers do not want the calves to get even a drop.

On an old-fashioned family farm, such as the one he grew up on, the author Jim Mason, a fifth-generation farmboy who is one of the world's leading authorities on farm animals, tells me that a newborn calf could stay with the cow for a couple of days—to get on its feet and to drink the colostrum, the mother's special first milk full of energy and antibodies, but even that is no longer permitted by agrifarms.

A dairy cow's male calf (primarily Holsteins and Jerseys) is taken from his mother and spends the remainder of his short life confined to a "veal crate" not much larger than himself. He never tastes his mother's milk and is deliberately made anemic by being fed a liquid diet with so little iron his meat will be white, a result much prized by consumers, often without any idea of how it is produced.

In the typical life of a cow or steer raised for beef (popular breeds include Hereford and Charolais), calves are weaned at six to ten months of age, live three to five months on the range, spend four to five months being fattened in a feedlot, and are typically slaughtered at fifteen to twenty months. Considering that their average lifespan is nine to twelve years, these animals live for only a brief fraction of the time they were meant to live. Joyce D'Silva tells me of one Irish cow she knew who died when she was 39! Nobody knows how long cows might live under natural conditions, but in domestication only one in one hundred thousand passes her nineteenth birthday. No cows or steers today live that long unless they are on a sanctuary.

Beef cattle—that is, cattle raised to be eaten—are sometimes sold at auction or, when they are taken to a feedlot, or to a more intensive indoor system, they are fattened up on a steady diet of rich feed and hormone implants. Jim Mason has seen up close what happens to cattle. "Think," he told me, "of a farm like my family's in Missouri. Here a few bulls are able to run with a herd of cows. They have it the best of all farmed animals today (except for the dehorning, castrating, branding, and other humiliations). All they have to do is roam around and eat, sleep, drink, and propagate. The calves are born and stay with their mothers for months until natural weaning time. The herd is together like a traditional cattle herd." But while it might be a good life for the lucky few, most of these animals are destined to die long before their natural life span. When the weaned calves get big enough, they are rounded up and sent to auction. These are called "cow-calf" operations, or, out West, ranches.

The cows used for milking are usually from a different breed (the most popular are the Guernsey, the Holstein, and the Jersey). Generally a calf is taken very young, for she is more easily fed from a bottle if she has never nursed from her mother. At about two years old, she is ready to have her first calf. (After she calves, a dairy heifer will become a dairy cow—"heifer" is only used until she has her first calf, after that she is called a cow.) The calf is taken away within forty-eight hours of birth, and the milk is then used exclusively for commercial purposes. The cow is rebred about three months after she calves. As long as she is pregnant, she gives milk, intended for her baby but taken by force by us. In the worst-case scenario, cows are intensively milked, most of the day and night by automatic machines worked by computers, and are exhausted after a few years, then sold for meat in repayment for their trouble. It is not a pretty life. On the family farm, rapidly becoming more and more rare, a cow is only milked twice a day, twelve hours apart. The cow is in charge, since she can let down or hold up her milk (not an option in the auto-mated farms) and is most likely to release her milk when she is calm and relaxed. Since she has four teats, it is possible to share the cow's milk with two other calves; they nurse one side while the farmer milks the other. A mature dairy cow will easily feed four calves at once—each calf, though, has its own preferred teat. A champion cow can give up to 110,000 glasses of milk a year!

Even Temple Grandin, the academic expert on killing cows "more humanely," has looked at a cow and admitted, "that's one sad, un-happy, upset cow. She wants her baby. Bellowing for it, hunting for it. It's like grieving, mourning—not much written about it. People don't like to allow them thoughts or feelings."[126]

This recognition of the feelings a cow has for her calf has af-fected some farmers deeply. Peter Roberts, the founder of Compassion in World Farming, was once a dairy farmer, and every dairy farmer eventually has to remove a calf from the mother. In this instance, he saw that "the motherhood bond was so strong, that it had to be broken with violence. It keeps you awake at night." But there was an even greater treachery. He noted "there are occasions when animals are terrified in a slaughterhouse. Mine didn't seem to be afraid because they trusted me. I broke that trust."[127] This is not an unusual experience. Stephen Clark, professor of philosophy at Liverpool University and author of the classic *The Moral Status of Animals*, was visiting Suffolk in 1972 with his wife, a historian who translated Porphyry's "On Abstinence from Animal Flesh." They were sleeping next door to a farm and were kept up at night by the complaints of the calves. The next day they found out that the calves were newly separated from their mothers. They said to one another, "We cannot go on financing this by eating meat and drinking milk," and both stopped doing so then and there.

All calves are separated from their mothers on all factory farms. John Avizienius, the senior scientific officer in the Farm Animals Department of the headquarters of the RSPCA in Great Britain tells me that he remembers one particular cow who appeared to be deeply affected by the separation from her calf for a period of at least six weeks. When the calf was first removed, she was in acute grief; she stood outside the pen where she had last seen her calf and bellowed for her offspring for hours. She would only move when forced to do so. Even after six weeks, the mother would gaze at the pen where she last saw her calf and sometimes wait momentarily outside of the pen. It was almost as if her spirit had been broken and all she could do was to make token gestures to see if her calf would still be there.

Shakespeare speaks of the plight of the calf in *King Henry the Sixth*:

> Thou never did'st them wrong, nor no man wrong:
> And, as the butcher takes away the Calf,
> And binds the wretch, and beats it when it strays,
> Bearing it to the bloody slaughterhouse;
> Even so, remorseless, have they borne him hence:
> And, as the Dam runs lowing up and down,
> Looking the way her harmless young one went,
> And can do naught but wail her darling's loss.

Is Shakespeare's portrayal of the cow wailing the loss of her beloved calf anthropomorphism? No one who has ever observed it doubts that a cow mourns the loss of her calf. Even farmers know that the longer the cow has known her calf, the more profound the grief, which is one reason most farmers remove the calf almost immediately after birth. Familiarity breeds love.

When we think about this love on the part of both animals, it becomes difficult to understand the human passion for veal, still very much alive in France and Italy among other countries. Most people, however, once they learn the details of the "life" of a veal calf, take their first step on the path to vegetarianism and foreswear veal for life. In the United States and in many parts of Europe, the sale of veal plummets once consumers understand the facts.

When given the opportunity, cows withdraw from the herd when they calve, isolating themselves as much as possible. The calf is then hidden, or "lies out" from the herd, and is only able to join when the "king" bull approves. But it is the rare and compassionate farmer who allows the cow to withdraw and give birth as she was meant to do.

The Torah teaches:

> If, along the road, you chance upon a bird's nest, in any tree or on the ground, with fledglings or eggs and the mother sitting over the fledglings or on the eggs, do not take the mother together with her young. Let the mother go, and take only the young, in order that you may fare well and have a long life. (Deuteronomy 22:6)

In his commentary on this Torah passage, the medieval commentator Nachmanides quotes Maimonides's theory that this commandment was given "in order to admonish us against killing the young within the mother's sight, for animals feel great distress under such circumstances. There is no difference between the distress of man and the distress of animals for their young, since the love of the mother and her tenderness to the children of her womb are not the result of reasoning but the result of feeling, which even animals experience."[128]

Our guilt about killing an animal in a particularly cruel way seems to go back to ancient times. There is an unusual passage in a Muslim holy book that reflects this: "The Holy Prophet said to a man who was sharpening his knife in the presence of the animal: 'Do you intend inflicting death on the animal twice—once by sharpening the knife within its sight, and once by cutting its throat?'"[129] Apart from the sentiment mentioned, this passage takes for granted that animals have consciousness and can anticipate (in this case with terror) the future. In any case, both the Jewish and the Islamic tradition have expressed similar observations based on a deep sense of moral outrage.

Everyone does not share the moral outrage, of course. I went to an annual agricultural show outside of Auckland, in New Zealand. There are contests for the best heifer, the best bull, and the best dairy cow, and I thought that by talking to the people who breed these animals, I might try out some of my ideas about their emotional lives. I spoke with a number of women who worked with these animals. "What do you see when you look at them?" I asked, hoping for some insight into their emotions. "I see good red meat," one of them told me, and her sister agreed. "What about their feelings?" I asked. "They don't have any," they agreed, and by now others listening added their views as well, all pretty much along these lines. "They are so even tempered," one woman told me. "They are always the same, they feel nothing." At that moment, we all heard a loud bellowing. I asked why the cows were making that noise. "Oh, it's nothing," the woman assured me, "just cows calling their calves." What did she mean, calling them? "Well, they've been separated and the calves are afraid and are calling for their mothers, and their mothers are afraid for their calves and are calling them, probably trying to reassure them." Here it was, from their own mouths, the same mouths that said these animals feel nothing, no fear, no pain at separation, no desire to comfort, no love for their child, no missing their mom.

I was led to some psychological or philosophical speculation.

People who love animals are often accused of indulging in the logical fallacies of sentimentalism or anthropomorphism. But now it sounded to me as if these people were suffering from another kind of fallacy, a psychological one known as *confirmation bias*, which involves only taking into account evidence that confirms a belief already held and ignoring or dismissing evidence that disproves that same belief. Against all the evidence, even that provided by their own eyes, these people convince themselves that animals feel nothing at all.

This has not always been the standard view. Francis Galton, a cousin of Charles Darwin who lived for a year among the cattle cultures of western South Africa in the 1850s, was fascinated by cows: "The habits of the animals strongly attracted my curiosity. The better I understood them, the more complex and worthy of study did their minds appear to me." He was particularly struck by how detached cattle seemed from one another, how self-absorbed. But perhaps this was the bias of the spectator, for Galton observed that when one of the animals was accidentally separated from the herd, "he exhibits every kind of mental agony; he strives with all his might to get back in again, and when he succeeds, he plunges into the middle, to bathe his whole body with the comfort of closest companionship."[130]

Some early writers recognized that not only did cattle have interesting minds, it was also possible to form a relationship with them. The earliest account we have of cattle in English is by William Youatt, who begins his book about cows by saying that "cattle are like most other animals, the creatures of education and circumstances," and later says, "He has become the slave of man, without acquiring the privilege of being his friend." But cattle can be "warmed with a degree of human affection." The most affecting passage in the book appears at the beginning, an account of a traveler in Colombia:

> I was suddenly aroused by a most terrific noise, a mixture of loud roaring and deep moans, which had the most appalling effect at so late an hour. I immediately went out, attended by the Indians, when I found close to the ranch, a large herd of bullocks collected from the surrounding country; they had encompassed the spot where a bullock had been killed [butchered] in the morning, and they appeared to be in the greatest state of grief and rage: they roared, they moaned, they tore the ground with their feet, and bellowed the most hideous chorus that can be imagined, and it was with the greatest difficulty they could be driven away by men and dogs. Since then, I have observed the same scene by daylight, and seen large tears rolling down their cheeks. Is it instinct merely, or does something nearer to reason tell them by the blood, that one of their companions has been butchered? I certainly never again wish to view so painful a sight:—they actually appeared to be reproaching us.[131]

This is what so many scientists today dismiss as a "mere anecdote." Cows don't weep real tears, they explain. Probably not. But can anyone doubt that they mourn and grieve and love their children as much as we love ours? Who are we to dismiss with human arrogance the depth or importance of these feelings?

Humans are sociable and need others of their species to thrive, as do cows. After all, cattle are basically herd animals. This means that they are sociable with each other and potentially with human beings. No doubt this is the reason we were able to domesticate wild cattle; humans could take advantage of their instinct to cooperate, to be friendly, and to find a place in a hierarchy, even an alien one. Although the ancestors of our modern cows are now nearly extinct in the wild, we have a fairly good sense of their lives, and they are not different from domesticated cattle today. Dr. Marthe Kiley-Worthington, an expert on cattle, points out that there have probably been few changes in the social organization, communication systems, and behavior from the wild ancestral to modern domestic cows.[132] Evolution has prepared a mother cow to behave in certain ways toward her young. Everything is designed to protect the vulnerable calf. Cattle hide their young, and in hider species, the calves have practically no scent for the first few days of life to reduce the risk of attracting predators. When a mother cow cannot lick her calf (to enhance this scentlessness), cannot feed the calf, and cannot be in her presence night and day, it produces a mental and physiological stress that is perhaps only understood by women who lose their children at birth.

The cow/calf relationship has even influenced the size and position of their eyes. The eye of the cow is very large, allowing in a great deal of light. The image entering the cow's eye is about three times brighter than the image in a human eye. This physiological advantage no doubt helps the cow keep track of her calf. The cow's eyes are positioned on the side of her head to give a wide field of vision (the human visual field is little more than half that of a cow), so she is able to see any predator threatening her calf. Cows often recognize their calves by coat color even from a distance. Smell also plays a role. Cows have an advanced sense of smell; they are able to smell sodium from up to six miles away, so they know when their calf is present or how far away she is. They can probably also smell when she is in danger, even when no longer in sight.

As for the bull, he too has been designed to protect his cows and calves. It has been suggested that the retina of a bull is normally attuned to the green of green pastures, and so when he lifts up his head, it is usually to detect danger. He is then inclined to "see red." It is clear that certain bright colors, ranging from yellow to red, sometimes cause bulls considerable irritation, perhaps because they alert him to a change in the expectable environment, hence represent threat to his family. The human belief that bulls hate the color red could be related to red as the color of blood. Little study has been devoted to the psychology of the bull in a ring, so we really do not know what he may be feeling. But a sense that he has been separated from his family and is not able to protect them would explain the panic and rage that ensue.

Almost nothing is known about how cows communicate with one another or about their calls—there are at least six and probably many more. Most of these calls involve the cow-calf relationship. Robert Schloeth, a German field zoologist who observed semiwild cattle in the Camargue in France for over two thousand hours, noticed something that had been overlooked by everyone else: calves have signals to let other calves know that they are about to commence play and that everything that happens after this signal is meant to be so interpreted. What would otherwise be deemed aggressive or sexual here only means play. They use a special call, one not heard in other circumstances, and a special run. He said: "They curl their wooly tails and wave them on one side only, which makes them readily visible as a signal."[133]

In Indian classical literature, as one might expect, cows are treated well. The heavenly cow Surabhi, the mother of all cows, wept and lamented when just two (out of billions!) of her children were exhausted and beaten by a plowman. And elsewhere in the epic the *Ramayana*, Kausalya is said to have cried out "like a cow seeing her calf bound and dragged away."[134] Gandhi once said, "Cow protection to me is one of the most wonderful phenomena in human evolution. It takes the human being beyond his species. The cow to me means the entire subhuman world. Man through the cow is enjoined to realize his identity with all that lives."[135]

Recently I visited a Norwegian graduate student in Cambridge, Kristin Hagen, who was studying the behavior of heifers—female cows who have not yet had a calf—and their emotional reactions to learning and individual recognition. We went out into the field to watch them, and after a while the cows began to play. "Watch," Hagen told me, "what they do when we begin to run." We ran. As we did so, the young cows ran through the fields with us, jumping and cavorting like children. They kicked up their heels, ran fast, then slowed down, kept us in their sights at all times, and even seemed to be imitating us. In any event, it was clear they were enjoying themselves. Looking into the eyes of two cows standing next to me, I could see that one was wary, the other friendly. I was quite certain of these cows' distinct personalities. But if we look into the eyes of a cow, how do we, in fact, know what we see there? How can we say, "I saw love there; I saw sadness"? Can we really read these emotions from the eyes themselves?

Darwin raises a similar question in a passage from *The Descent of Man*:

> It is often difficult to judge whether animals have any feeling for the sufferings of others of their kind. Who can say what cows feel, when they surround and stare intently on a dying or dead companion?[136]

It is difficult to get inside the mind of another animal, Darwin acknowledges, but at the same time he seems to suggest that he believes cows could indeed feel sadness. He goes on to qualify his statement, referring to a remark by the French naturalist J. C. Houzeau that cows "feel no pity." Houzeau lived for many years in the United States, sleeping by himself out on the wild prairies before becoming director of the Royal Observatory in Brussels and a famous astronomer.[137] Of his close observations of cattle he wrote:

> I myself have seen many instances of cows stuck in the mud along the banks of natural lakes in Texas. I have carefully examined the expression and the behavior of the cattle who came to the same lake to drink just a few steps away from these victims. One day in the province of Comal I came across a tiny water hole, on the bank of which a cow was immersed up to her chest in the mud. No doubt she had tried very hard to get out; but she was now completely exhausted: she was barely moving, and even the few feeble gestures she made were few and far between. Her eyes were beginning to

dim, her nose was covered in flies, and her death was imminent. In my presence other cows arrived right next to her, as she lay stuck in the mud. They paid not the slightest attention to this unfortunate animal, and far from making any effort to help her; they seemed to refuse her even their pity.

This is without doubt the passage Darwin had in mind.[138] As if responding to a critic, Houzeau then notes: "But as soon as we reflect on these situations and ask ourselves how, precisely, a cow who has not fallen could possibly help one who has lost her balance, we see that in fact, cows lack the means for assisting one another. The species has no prehensile organs: feet, tail and mouth are not made to seize, pick up, or pull. The forehead and the horns could only be used underneath another animal, and not without injuring the very animal they are trying to save. The indifference of the bystander is really but the result of physical impotence."

If one agrees with this passage, it is then impossible to know whether the cow feels anything on these occasions. The only way we have to evaluate pity or sympathy in such circumstances is to observe behavior, and in its absence, we really cannot know what is going on inside the mind of the impotent animal.

Donald Broom, professor of Animal Welfare at the University of Cambridge, has observed that only animals who are accustomed to being helped cry out in pain when they are injured. This is the case for dogs, pigs, and humans. In animals such as cows (or sheep), however, where no help can be expected, crying out would only attract the attention of a predator. Broom says there is no question that these animals who do not cry out cannot be said to feel pain any the less.[139] But in the absence of communal signs, such as physical gestures or sounds, humans are simply not equipped to understand animal emotions. This does not mean they are not there. For example, Professor John Webster has written convincingly about fear and its expression in cows:

> Pigs, puppies and rabbits are more likely to cry out when in pain and fear than, say cattle or wild ruminants. . . . The puppy that yelps is trying to attract help . . . The wild scream of the rabbit or pig may unnerve its predator or alert its siblings to the existence of danger. Many of the species that scream the most are those born in litters rather than as singletons. If a rabbit screams when it is captured its siblings may escape . . . the deer calf on the African plain is on its own. If it cries, limps, or displays other obvious signs of distress it becomes the individual marked out by the lion as easy prey. It is likely that the domestic cow and sheep display similarly Spartan behaviours, i.e., they try not to reveal how much it hurts.[140]

Webster has a very strong sense of how much pain cows have to endure basically in silence and with a complete lack of comprehension on our part. Because cows are now genetically selected to produce approximately ten times as much milk as their calves would suckle from them, they suffer from both mastitis (experienced by 35 percent of all cows) and damage to their feet, since, with the udder so full, they must walk in a totally unnatural manner. Lameness is now so common as to affect some 60 percent of all cows; their pain is beyond human description. Webster tried to illustrate it for me: "Imagine that you caught all your fingers of both hands in a doorjamb, hard. And then you had to walk on your fingertips. Can you imagine what that would feel like? So when you see a cow hesitating to put one foot in front of the other, you can be sure that she is feeling excruciating pain." He pointed out to me that there is new evidence to show that chronically lame cows display hyperalgesia, or increased sensitivity to pain. In this regard they respond as we do. They and we do not adapt to chronic pain; the sensation gets worse with time.

As with the Tamworth Two pigs, we instinctively feel sympathy for an animal that escapes its "destiny." In 1995, in the small town of Hopkinton, Massachusetts, a five-year-old, black-and-white, 1,400-pound Holstein was in a slaughterhouse about to be stunned then skinned, when, like a reindeer, she suddenly leapt over a five-foot fence. The slaughterhouse employee who saw it happen said he could not believe his eyes: "She just saw what was coming. The cow escaped into the woods where she was thought to have joined a herd of deer. She somehow acquired the name of Emily. People spotted her in the woods, and her reputation as an unusually canny cow grew, as did sympathy for her escape. For the five weeks she was free, the slaughterhouse determined to catch her and kill her. But a group of town residents, hoping to save her life, persuaded the owner to sell her to them for one dollar. Now the only problem was catching her: she was wary of people and kept to the woods. The police and animal control officers tried to catch her but said she was too quick and wily for them. One of the new owners, Meg Randa, followed her into the woods, sat down, and did not move for four hours. Eventually, Emily, curious, came over and accepted food from a bucket of grain that Meg's husband, Lewis, was holding. For four days the couple stayed nearby, slowly gaining the cow's trust. At last, after her forty days in the desert, Emily was willing to walk into the trailer they parked there for this eventuality. She was shaking when she entered but was taken to the Peace Abbey, a school and shelter for animals in Sherborn that was started by the Randas after Mother Teresa visited them in 1988. The Randas and their three children had Christmas dinner in the barn with Emily and two horses, a goat, several rabbits, and some cats and dogs. The meal was vegetarian. Emily is now their "spokescow" against the use of red meat. Lewis says that people "can listen to Emily or they can listen to their cardiologist."

Kimberly Muncaster, the manager of the New Zealand branch of the World Society for the Protection of Animals (WSPA), an international animal welfare organization, told me recently how she stopped eating beef. A friend told her that she passed a slaughterhouse every morning on her way to work (in Perth, Australia), and she noticed the cows lined up in the preslaughter pen from where they could see their companions being killed. They were trembling—they could hardly stand up they were shaking so badly. They were absolutely terrified. When Muncaster heard this she realized she could never eat an animal whose end was so miserable. She became a vegetarian from that day on.

Domestic cattle, except those found in Southeast Asia, are all descended from a single wild species called the aurochs—the last of which was killed in a Polish park in 1627. The bulls were large, up to six-and-a-half feet at the shoulder, and they were often equipped with very long horns. These were formidable animals, swift, strong, courageous, and with no affinity for man. Wild cattle in India, known as gaur, have some resemblance to these original cattle.

American zoologist George Schaller once observed a tigress walking, followed at a distance of 150 feet by eight gaur. Tigers seldom attack adult gaur because of the gaur's tremendous strength and courage. These ancestors of our domestic cattle have actually been "reconstituted" by Heinz Heck, the director of the Munich Zoo, and his brother Lutz. They crossed Hungarian steppe cattle with Scottish Highland cattle, Alpine breeds, Corsicans, and others, on the theory that the various characteristics of the aurochsen (plural of the singular aurochs) are still to be found in different contemporary breeds. It worked, and by 1951, forty of these reconstituted aurochsen were allowed to run wild in Bialowieza in Poland.[141] They match almost exactly pictures of aurochsen from the French cave paintings in Lascaux and the four hundred even older paintings in the Ardeche, which have been carbon-dated to 31,000 B.C.

Most authorities estimate that cattle were first domesticated about eight thousand years ago in the Middle East. Today, worldwide there are approximately 1.2 billion cattle. Buffaloes—larger and stronger than ordinary cattle—can be milked, ridden, and harnessed to the plough or to carts and wagons, and can work up to their noses in floodwater, which is why they are so important in rice-growing lands. Buffaloes were first domesticated in India about five thousand years ago. As Edward Hyams, a writer about domestication, points out, farmers in Italy didn't use buffaloes until about A.D. 800; it took them nearly four thousand years to get there from India.[142]

Nobody knows for certain why cattle were first domesticated. Since humans ate the meat of wild cattle, it has been assumed that domesticating these animals would give them a continual supply of meat and hides, but this is far from certain. It was not for milk, since most of the world is lactose intolerant (especially the population of central Africa and eastern Asia, where there is no tradition of milking). When such Roman writers as Columella and Varro discuss agriculture, they do not mention cattle except as draught or sacrificial animals. The Romans did not drink milk and, as the expert on domestication Juliet Clutton-Brock points out, "this is reflected in the physiology of their descendants today." In countries where milk is routinely drunk, people are larger, but this may not be a health advantage, for in many animals artificially induced larger size leads to a shorter life span. One cannot think of any possible advantages cows could derive from giving up their freedom and choosing domestication. They did not need us to protect them from enemies, as powerful as they were; no animal could kill an adult auroch, and circling would protect the young.

Dr. Marthe Kiley-Worthington ran a sanctuary at her Little Ash Eco-Farm in Dartmoor National Park in Devon, southwest England, part of the ancient uncultivated moorlands. Surrounded by rather tame laughing doves, I spoke with her about cows. "Cows," she told me, "have changed my life by sharing with me a different way of life in the wild. They don't complain; they are not manipulators of the world. They cope. They feel but just don't make a fuss. They think positively and are incredibly tolerant: 'Well, here I am. How can I make the most of it?' They have such serenity. And remember," Kiley-Worthington told me, "they have a huge brain." True, if they were just instinctive beings, they would not need such a brain. It has more folds and convolutions than that of a dog and nearly as many as the human brain. A brain requires a huge amount of nutrition to develop, so why did the cow's large brain evolve? Nobody knows.

I watched while Kiley-Worthington asked Rambler the cow to lift her left leg. She did so. Then Rambler was asked to lift her right leg. Again she obeyed. Then she was told to go over to the fence where there were different colored squares and to put her nose on the blue one. She did it. Then the yellow one. Success again. Then she was asked to paw the ground. She pawed. I was amazed.

We talked about the dreams of cows. Cows dream, of that there is no doubt.[143] Nobody, of course, has ever cracked the code of precisely what they dream about or what they think about when they seem lost in contemplation while they quietly chew their cud. Given, though, that cows are often forced to observe their companions or children being taken away, I wondered if they might also have nightmares. I had heard from other people who lived with cows that they had seen them look agitated in their sleep. What disturbs a cow, I wondered? Could they be afraid we are going to kill them?

Some horse experts think that one reason horses are shy around some humans is that they fear precisely this.[144] Horses, at least, can outrun many of their enemies, but cows do not have this protection and must feel quite vulnerable. They depend entirely on the protection of male bulls, who are no longer around the herd. This in itself could produce anxiety. Kiley-Worthington agrees, but she also believes that the physical problems resulting from the way cows are now bred produces enormous stress on them. "Think," she told me, "of the stress of the cows now bred for double-muscling, with the huge backsides, for example, the Belgian blues. The bulls are so enormous they can barely walk. Their feet are simply not designed to carry such weight. The extra weight means that females cannot have natural birth and require caesarean, which are done only with a local anesthetic, and sometimes not even with that. Even though they don't complain, their heart rate reveals that they are in intense pain and are suffering." Just because cows don't show their feelings doesn't mean they don't have them.

Recently I went to visit Jamie Brown and his wife, Michele, who live with eighteen cows and three bulls on seventy acres of native bush, rain forest, and lush meadows on the Coromandel Peninsula in New Zealand. James is getting a diploma in homeopathic animal therapy. Michele is running their retreat, a true oasis of beauty and peace which they call Vida. We walked out to their pasture, sat down on the lush grass, and quietly waited. The cows were far in the distance when we arrived, but slowly, almost imperceptibly, they approached. As they inched ever nearer, they began to call. Clearly these sounds were significant, not wailing, or just vocalizing, as too many scientists would say, but communicating. They obviously had something quite specific in mind, and the only problem seemed to be ours, that we could not understand what they meant. Soon the cows had surrounded us and were staring. They seemed as curious about us as we were about them. The large bulls were muscular Belgian blues, gentle and affectionate. Jamie told me to look at how delicate their eyelashes were, something I had not noticed, but now that I looked, they were indeed really quite lovely, long and silky.

I asked Jamie what he had learned from cows to make him seek out their company in this way. So few people feel close to cows, though sitting here in the still of the meadow, near the forest, watched by thirty-six soft eyes, I wondered why that was so. "I learned about love," Jamie said. Daisy, a cow he was particularly fond of, had jumped four fences, found a bull she liked, jumped back over the four fences, and returned pregnant. Jamie thought she would make a terrific mom. He was looking forward to seeing the cows with their calves, for he knew that they made attentive, gentle mothers. Rarely are they given the chance to be so under most farming conditions, since their calves are taken away immediately. Not these cows. Nor did they have to worry about a shortened life—no butcher was waiting for them. They were there because Jamie and Michelle loved having them around. Being with them in the grass, surrounded by their low mooing, feeling their damp breath on my hands created a wonderful feeling of peace. I could have sat there all day. The retreat Jamie and Michele founded is intended to be a place of healing, and the presence of the cows only intensified the sense of time slowing down, offering the possibility of another, more compassionate kind of life, one where serenity has the highest value. In such a place, just learning to be still with cows could be an exercise worthy of the best Zen meditation master!

Surely no animal was born for just a quiet life on a hillside followed by a quick slaughter? Cows were not put on this earth to give us their flesh, nor even their wisdom. However, it makes more sense to take the latter rather than the former.

Maybe it was the setting. Fall, a certain dark melancholy was in the air. We were driving in the South Island of New Zealand, just about to cross a river, when I felt rather than saw the presence of some large animals. I was being stared at. I turned around in the car seat, and sure enough, there was a feedlot, within it about two hundred cows and bulls. They were beautiful animals, all black. Against the darkening sky, the leaves changing color, the dark blue river, it was as if I were looking at a painting. They were healthy, strong animals, clearly well cared for. Yet here they were in a small enclosure on this beautiful fall day, the sun about to set. They were not doing what cows normally do. They were not grazing. They were staring at me.

I stopped the car, turned around at the bridge, and drove back. I parked in front of the lot and as many of the animals as could crowd around came up to the fence of their feedlot, just staring at the four of us, my wife, my two small children, and me. The silence was eerie. Their look seemed full of purpose. What were they looking at? What did they see? Any answer, I realized, was pure speculation— speculation of the kind that scientists hate. So yes, while my ideas about the cows' thoughts were nothing but speculation, that does not mean my speculation does not reveal something.

What I imagined was that the cows were wondering why they were there, singled out for death. What kind of world was it that allowed my family and me, cozy and safe in our new car, to simply drive by, whereas they and their family were destined to be driven away in a large truck and murdered for their meat? I left, watching them watching me leave, following my departure with their sad eyes. How could I do nothing for them? Guilt welled up in me. There was nothing I could do for them. But that is not entirely true. I could write these words and these words could be read by people like you who might feel what I felt. One day, far in the future, people will marvel that we took the lives of these gentle and beautiful animals to satisfy our greed and gluttony. And one day a family much like mine will drive by and cows much like these will be grazing on a hillside, and those cows will be admired rather than eaten by humans.

FIVE

Like Ducks
to Water

Ducks and geese are the only farm animals we have all had the chance to observe and enjoy in the wild. Our desire to be around them and to watch them begins in earliest childhood. Our son Ilan, who is six, was always excited to visit a duck pond, and now Manu, our seventeen-month-old, is equally thrilled. One of Manu's first words was "duck," uttered as he pointed excitedly at the gracefully moving creature he saw gliding across the water. Most small children derive great pleasure from feeding ducks. Nor does this pleasure disappear when we become adults. It is one activity that every grandparent can enjoy. Perhaps it awakens in us nostalgia for the loss of childhood innocence. The nineteenth-century Irish poet William Allingham wrote a short poem, often anthologized, in which this loss is poignantly, if a bit cryptically, memorialized:

Four ducks on a pond,
A grass-bank beyond,
A blue sky of spring,
White clouds on the wing:
What a little thing

To remember for years—
To remember with tears!

When I was six years old, we lived in an urban California neighborhood. It is unclear to me how I came to have ducks in our backyard there, but have them I did. I remember being sent home from school on the first day of first grade because I had arrived accompanied by two ducks who would not leave my side. I was upset when I was forced to lock the gate of our backyard and hear them left behind, pleading for me not to leave them. The meaning of their calls could not have been more clear.

When we moved a few years later into a home in the Valley, I was present when our ducks' eggs hatched, and I was the first living being the ducklings saw. They imprinted on me and wanted to go wherever I went. I found it flattering. The problem was that we had a swimming pool, and the ducks loved it when I went swimming, because in they went, too. That annoyed my parents no end. Of course I did not understand what imprinting was; I thought my ducks just loved me. Well, perhaps they did. After all, what is love but a kind of psychic imprinting, the conditions for which elude our understanding?

Fernando Yusingco is a spry, seventy-three-year-old Philippine man who lives on a twelve-acre duck farm an hour north of Auckland. Yusingco has been raising ducks here for the last fifteen years. Before that, he had ducks in the Philippines. He has about 500 ducks on his farm, 100 of whom are layers—that is, they produce fertile eggs from which baby ducks are hatched. These ducks, a mixture of Pekin (bright white) and Cayuga (dark brown), are like almost all domestic ducks: they cannot fly. But they have a huge natural pond to swim about in and some five acres to wander, including a natural stream (which they seldom use; for some reason they do not like moving water).

I recently had coffee with Yusingco in his living room and watched the colorful scene outside the window. "I thought you said they couldn't fly," I said, watching a large duck circle the pond and land on the water with a graceful dive.

"That's a wild Canada goose, and there's a mallard coming down, too," Yusingco explained. There are eight species of wild duck native to New Zealand, but the mallard is not native; it was introduced by Europeans only a bit more than a hundred years ago. The other species include the paradise duck, the blue duck, and the gray teal. Why would they come to Yusingco's pond?

"They seem to like the company. I think they are curious about all the ducks on the pond and come to investigate. Sometimes I'll look out the window, and there will be hundreds of visiting wild ducks of every sort. I love it." As we sat and talked, it became clear to me that Yusingco was no ordinary duck farmer. He had been a labor organizer and a political foe of Marcos who had spent years in jail in the Philippines. He was glad to have a peaceful, quiet life on his duck farm far from his former political life. He sold live ducks on the weekends (twenty to fifty of them), for eating. He sold them to people in the Asian community of greater Auckland, including Indians who liked to make duck curry with the older ducks (three years old). He also sold their eggs to the Chinese, after curing them in salt for several months. He ate duck omelets every week, but, strangely, no duck.

"Why not?" I asked, and watched him shuffle his feet.

"Truth is, I kind of like ducks, and I don't seem to have it in me to kill them myself." To give them out for slaughter felt to him a bit like betrayal. What an odd and appropriate word for a duck farmer to use!

"So you like ducks?" I asked.

"Actually, I love them," he admitted. "I love how peaceful they are. They never fight. Or almost never. Sometimes there is a brief quarrel, if one ducks happens to be sitting on a nest, and another duck wants to get by. They don't fight with each other, and they don't fight with the visiting wild ducks either."

On Yusingco farm there is one drake for every four or five ducks. Charles Darwin was the first to point out that ducks in the wild are monogamous—at least for the season—but when they are domesticated, they become polygamous.[145] So I asked Yusingco if the drakes were entirely happy to mate promiscuously.

Here is where any good observer can correct even the great Darwin. "Well, actually," Yusingco told me, "I have been surprised by what I sometimes see. There are times when a drake seems to prefer one or two females to all the others. He seems reluctant to couple with any duck but his chosen one or two." Domestication may have made drakes promiscuous, but they still seem to have a pull in the other direction, a faint memory, perhaps, of the joys of having a partner! But the drakes will also force themselves upon a female, and when they do, that female has to be isolated to protect her. I have never understood this and wonder just how commonly it occurs. Of course we have nothing like statistics, but apparently it happens in the wild and is not just an artifact of human intervention.

While Yusingco clearly likes his ducks, they do not return the favor. They seem leery of him. He doesn't take it personally; they are suspicious of all humans, he says. In fact, Yusingco thinks that ducks have as their motto: "Never trust a human being." I can understand. After all, even Yusingco, who treats the ducks on his farm so humanely, sells them to be eaten.

He sells most of the ducklings when they are a mere thirteen weeks old. After that, they don't grow any larger, so there is really no (economic) point in keeping them beyond that age, Yusingco explains. Since even the layers don't lay as much once they reach their peak of 250 eggs a year, at two years, they too are sold off—poor gratitude for their 500 eggs, it seems to me. Think, too, of how artificial their laying is: in the wild, a mallard lays from five to ten eggs, twice, or at most, three times a year. She broods them, and then they hatch. On farms, the eggs are candled, that is, a light is shined upon them revealing whether there is an embryo inside; if there isn't, the egg is destroyed; if there is, it is incubated artificially. Needless to say, the enormous increase in egg laying has caused serious physical problems for the duck, including diseases of the female reproductive organs, which were not designed for this volume of egg laying. In the wild, the duck stays with her ducklings several months. On most farms, ducklings are killed long before they would even have left their mother.

The tragedy is that ducks, like other animals raised in America for food, are excluded from the federal Animal Welfare Act. There are no standards set by the U.S. government in terms of how these animals are housed, fed, or treated on farms or even in somebody's backyard, as long as the claim is that they are being raised for food. Anything that is "customary" is permitted. Farmers, therefore, have few incentives to treat their animals in a humane manner unless they stand to profit from so doing.

Here we come to the crux of the matter: if you are trying to make money off the backs of ducks or any other farmed animal, you may treat them with a kind of humanity, as Yusingco clearly does, but in the end, you are still selling them short, because you are selling them for food. In other words, as long as our purpose in having farm animals is to kill them and exploit them for their eggs, their milk, their coats, or any other part of their body, we cannot live with them in any kind of healthy relationship any more than we could with another human being whom we exploit. Knowing how passionately opposed Yusingco was to the exploitation of poor people, I was most curious to learn his views about animal exploitation. When asked, this was his reply: "In the order of creation, animals are meant for human survival. That is why they are there. Unless one believes that they are there only for companionship, to share the earth with us, but not to provide us with food as well, then one must take the course that is common today, namely, animals are also for food. The common practice of the majority of people cannot be barbaric. Maybe someday the world will turn vegetarian and we will stop killing animals for meat. But is there any moral ground not to kill animals for meat? If their *raison d'être* is to provide food for humans, am I exploiting them if I eat them?" Obviously, my answer would be yes.

I had a most extraordinary experience while I was at Yusingco's farm. The ducklings who were just born the day before were all in a box. When we approached, they fled to a corner of the box, as far from us as they could get. Yusingco went out to see to some task while I stayed behind, fascinated by these bright yellow fluff-balls.

I had my hand in the box, and to my surprise, after a minute or two, one of the bolder ducklings came to investigate. He pushed against my finger, so I began stroking his back. I had no idea how soft newborn down feels on the back of a small duckling. I loved stroking him, and he loved being stroked! A moment later, perhaps emboldened by what they saw, two or three of the other ducklings waddled over to my hand and got the same treatment. They, too, enjoyed it. They did not run away from me; on the contrary, they jostled one another to get under my finger and feel the soothing touch. Soon all the ducklings were scrambling to be with me. With theirbright little black eyes, and their brilliant orange beaks, and their perfectly formed webbed feet, they were exquisite little creatures.

They stayed in their box a few days and then were moved to a small pen indoors. Yusingco did not mind my cooing over them (I think he secretly felt just like I did). As dusk fell, the rest of the ducks voluntarily left their pond and waddled over to a large barn where they spend every night. They did not need to be herded in; they evidently felt safer at night in the barn than in the pond. Yusingco told me that each evening there were always a few adult ducks (never the drakes, though) who would jump over the side of the pen in which the small ducklings were kept. They would fold their wings over as many as they could get underneath, and then sleep the night with them. Could these ducklings be the offspring of the ducks who spent the night? "Who knows?" admitted Yusingco. "It wouldn't entirely surprise me if they were." These ducklings had not imprinted on me, but they were clearly all too happy to hear a friendly voice and obviously liked the feeling of being stroked. Maybe they were just lonely for an adult protector, but they were overjoyed that I paid them some attention. It broke my heart to leave them, knowing their fate.

We have a tendency to insult the intelligence of domestic fowl. Even the late Ted Hughes, the English poet laureate and the preeminent poet of animals, described hens as "empty-headed," who think, in their domesticated bliss, "that the fox is a country superstition."[146] In fact, not only are fowl eminently aware of the movements of the fox, they will communicate their presence to other animals. Ducks in a pond who have caught sight of a fox on the bank will mass together, then swim parallel to the fox as he trots along the shore in what is an unmistakable visual signal to all other inhabitants of the pond: "Look out, the fox is here!" Scientists find this behavior, which *resembles* altruism, troubling and puzzling: How could a duck wish to warn other animals about a dangerous fox? What's in it for the duck who is chosen, or chooses, to be a sentinel? Is it just a warning, or are they mocking their enemy, secure in their knowledge that he cannot attack them? (No fox can swim as fast as a duck.) Is it a means of letting him know that he has been seen and there is no point in staying around, for tonight he will not have duck for dinner? These are among the secrets ducks keep.

Many people have happened upon a duck when she was with her ducklings, only to notice something strange: the duck suddenly seems to be injured. The ducklings scatter, and the duck begins what looks like a crazy dance, fascinating to us and evidently to other predators (of which we are just a larger version for a duck). We cannot help but follow the "injured" duck, forgetting her young. Suddenly, when we are lured farther and farther away from the ducklings, she is miraculously cured and flies off, emitting the equivalent of a sigh of relief: safe at last. We have just witnessed a display for which many birds, especially ducks, are famous, called the distraction display—or injury-feigning—acting as though their wings and backs were broken. The classic description, from a 1913 account, speaks of the birds "striking strange attitudes lying on the sand as though they had been wounded and fallen on the ground; others were floundering in the sand as if in pain; some were fluttering along with one wing stretched out limply, looking as though it were broken."[147] For anybody who witnesses this amazing trick, it is hard to believe that the mother bird is not conscious of what she is doing. Increasingly, ornithologists are prepared to admit this once hotly contested fact.

It seems that one reason we love ducks is that their calls are readily understandable. This is surprising, when you think about it—we do not generally understand all the vocal communication of our house companions, dogs and cats for example. Barking, which I have always thought is an attempt to imitate human speech (wild dogs do not bark) is still largely unintelligible. The desertion whistle of a duckling, on the other hand, is easily translated: "I am lost, come find me." In *King Solomon's Ring*, Nobel Prize winner Konrad Lorenz describes the lost call as "that penetrating piping of abandoned ducklings that we are accustomed simply to call crying."[148] Says the Australian writer Bryce Fraser, who raised many ducks, "I'm probably one of the few world authorities on duckling cheeps: I could compile a dictionary of duckling cheeps. I've heard them all: happy cheeps, sad cheeps, panicky cheeps, desperate cheeps, when-do-we-eat cheeps, where-are-you-mother cheeps, I'm-coming-mother cheeps. Most people believe ducklings simply cheep. But this isn't so."[149]

It is interesting that the slow quacks between adult ducks (indicating affection) derive from these infantile sounds. No doubt the slowness indicates that it is a comfort sound rather than a desertion sound. A desertion sound is difficult to misunderstand, the nonstop peeping is clearly from a bird in distress and almost sounds like help, help, help! A female's decrescendo call during fall and winter suggests that individuals may be trying to locate one another at the end of the season. If this is true, it may be that old pair-bonds are renewed more commonly than has been generally supposed for ducks.

I asked a group of friends how ducks were regarded in New Zealand. The answers were immediate and enthusiastic. "As birds to shoot at," came one response which I ignored, but then others told me how many people loved to attract ducks to their backyards. In fact, in New Zealand, I just learned, when somebody asks how you are doing, you answer, "A row of fluffy ducks," i.e., splendidly!

I was told of an elderly couple at Dairy Flat, just north of Auckland, who had a large duck pond in their yard. I was given their phone number, and soon we were on our way for a visit. As we approached their long driveway, we heard a shot, then another. You don't often hear gunshots in Auckland. When our hosts—Rex and Robyn Neary, an insurance executive and a retired librarian—

emerged from their house to greet us, I asked them what was happening.

"Today is the opening day of duck season," Rex said grimly. "Come with me," he added, "and you will see how the ducks respond." We walked over to the edge of his grass, and there was a large lily pond, teeming with ducks. The Nearys have two ponds with a creek and a swampy area in between, surrounded by native bush. The ducks seem to love it there, especially when elsewhere they are being shot at. "They know," he marveled, "I don't know how, but they know. It's the strangest thing, but every year, the night before duck season starts, our pond fills up." Fernando Yusingco had told me the same thing about his ponds during duck season: wild ducks knew it was a sanctuary. Ducks want to be safe. They know those shots are meant for them, but they must also know that there are places they can go where they will not be shot at. I wonder, does the word get passed around?

A man who hunted goose told me that geese, like ducks, are especially canny when it comes to the hunting season: "They are well accustomed to us hunters and are careful to drop out of sight the day before open season. Beats me how they know." He seemed to believe that because they had this uncanny knowledge, they were somehow fair game—as if being intelligent entitled you to be stalked and killed. Strange logic.

About eighty ducks live on the Nearys' pond now. They are wild mallard ducks, free to come and go as they wished, but this pond was clearly home for them. They have become quite tame. Some will even allow themselves to be stroked as they eat from your hand. A tame mallard is not very different from a wild mallard; in fact it is the identical bird, just with a different relation to humans. We are all familiar with ponds in parks where ducks of various kinds congregate and swim right up to you as you feed them. These, too, are wild ducks who have become tame.

Rex and Robyn Neary have lived in that peaceful setting for twelve years, and during that time the pond's duck population has been growing. For one thing, families tend to stay together. Once the female lays her eggs and is safely brooding them, however, the drakes leave. "I don't know where they go, exactly," Rex explained to me, "but they disappear for awhile." It is not cold enough for them to migrate, so either the female wants them to leave, or they need to be on their own. Either way, off they go in tight little male gangs, while the females stay behind to tend the eggs. When the ducklings hatch in early spring, the males return, but it's the female who takes care of them, swimming gracefully in the pond with the sight we are all so familiar with: the ducklings swimming behind her, imitating her every move. At the very back of their lake, lives a pair of paradise ducks. Paradise ducks mate for life, Rex told me, and if one is shot during duck hunting season, the other calls mournfully for days on end. "Sometimes we see the rare endangered New Zealand blue duck. I believe they also mate for life."

Most ducks do not mate for life, although some species are remarkably faithful to one another. Theodore Xenophon Barber, in his wonderful book *The Human Nature of Birds*, writes:

Reporting on a pair of mated mandarin ducks living in his large aviary, an English gentleman testified that, when the male drake was taken away, the female duck "displayed the strongest marks of despair at her bereavement, retiring into a corner, and altogether neglecting food and drink, as well as the care of her person." When her original mate was subsequently recovered and returned to the aviary, "the most extravagant demonstrations of joy were displayed by the fond couple."[150]

I heard a report of a drake who was particularly attentive to a duck, even after the mating season was over, unusual for male ducks. Dr. Arthur Peterson, in DeBary, Florida, had a large lake on his property and noticed this unusual behavior. When he went to investigate, he had to slip a net over the head of the duck to examine her. It was only then that he realized she was entirely blind! As soon as he released her, her guide returned and immediately approached her, making reassuring sounds, and leading her back to the lake and his constant vigil. He was a seeing-eye duck.[151]

It is true, alas, that drakes of some species (mallards, in particular), rape the female ducks (possibly this is not the right word— forced copulation might be a better term; we really know very little about what it might mean to a duck). Fernando Yusingco, as we saw above, observed this on his farm. These male sociopaths are known to form a gang of thugs, coming together specifically in order to have a better chance at raping a duck. These gangs will even attempt to rape a duck with ducklings or with her mate right next to her. (Her mate will do all he can to prevent the rape but is often unsuccessful.) It is not uncommon for the female to drown in the process. What is not known is whether *all* drakes do this, or only psychopathic ones. The strange thing is that after the males dismount, they give the same kind of triumph display that couples engage in after successful copulation. The females of course *never* give the display after they are raped. Is it possible that the males *know* they are doing wrong? After all, only *some* drakes perform the display.[152]

Males are definitely less picky: when mallard drakes are exposed over the first fifty days after hatching to a duckling or foster mother of another species, many will choose mates belonging to the other species when given a free choice. When female mallards, on the other hand, are raised with members of other species, they do not become imprinted but tend to mate only with their own species.[153]

Almost all animals have some version of imprinting, but no animal demonstrates it as conclusively as a duck or a goose. When ducklings or goslings hatch from the egg, they are attracted to the first thing they see—normally, their duck mother. Since ducks and geese are relatively independent at birth, they are ready to swim with their mother within hours of being born. Many ducks swim about with their young on their backs, like swans—a lovely sight. If for some odd reason (and under natural conditions it must happen rarely), the ducklings are separated from their mother when they emerge from the egg, they follow whomever else is there. Sometimes that is us, as I saw as a child when my ducks believed I was their mother.

My friend Matthew Scully, who writes speeches for President Bush, was surprised to see a mother and her ducklings come walking right by the White House in downtown Washington, and he wondered why they would nest in a city when there's miles and miles of river around the city. I think the answer, surprising as it is, lies in the ability of ducks to recognize that there are places where humans protect them from both their natural predators, other animals, and from their unnatural predators, namely us. Nobody would ever shoot a mother duck walking across a busy city street, and this should perhaps make us realize that there is *never* a reason to shoot a duck, mother, father, or baby!

The term "waterfowl" (in America) or "wildfowl" (in Britain) is applied to some 142 species of ducks, geese, and swans. They all belong to the same Anatidae family (from the Latin *anas*, to swim) and have much in common. Their bodies are designed for aquatic life, with webbed feet, short legs, long necks, and beaks with horny plates that serve as filters. They all have thick, waterproof plumage with a dense coat of insulating down underneath. Their young leave the nest shortly after hatching. (Ducklings reach flight stage between seven and nine weeks old. It is interesting that even if their mother is killed, the ducklings can travel up to a mile to find water on their own.) Covered in waterproof down, babies can swim and feed themselves hours after birth. Since these characteristics are common to all the species, they are united into a single family (with many subfamilies of course). Many are common birds known to everyone, such as the mallard duck. Others, like the Hawaiian goose, or nene, can usually only be seen in zoos, since there are but a few hundred alive.

The well-known British ornithologist Peter Scott was not the first person to realize that swans had different face patterns. Konrad Lorenz knew all his wild geese at first glance, much the way a good shepherd knows his sheep. It is possible, too, that ducks and geese recognize each other from signature whistles. Humans can definitely recognize a specific duck from his or her voice.

I have had difficult conversations with scientists about ducks and geese. Perhaps because they were the first animals to have demonstrated with such clarity the imprinting mechanism, many scientists think of them as purely driven by instinct. They have, say many scientists, nothing but what Konrad Lorenz in 1950 called "fixed action patterns," that is, they inherit a number of inborn movements, and these are rigidly organized. It does not matter whether these movements are irrational or harm the goose or duck, they *have* to come into play; the birds have no choice in the matter.

"Look," said one exasperated scientist I tired to argue the point with (he asked not to be named), "an adult duck will drive a strange duckling from its nesting area, but if the little guy succeeds in touching the duck's breast, then a different instinctive response is thrown into gear: she will accept the duckling as one of her own and brood it with great care. But if the duck is driven away, dripping with blood, and I have seen this happen, and a drake comes along and begins to torture the little one, the duck will immediately come to its rescue, but if and only if the duck hears the distress calls."

"What's wrong with that?" I wonder.

"Well, don't you see, the duck is simply a creature of the moment, responding to strangeness, to breast contact, to distress calls, all with sublime impartiality. She doesn't care a fig about the duckling itself; she is an automaton, primed to respond in a certain way. She is not cruel when she rejects the duckling and she is not compassionate when she rushes to its defense. She is a creature of pure instinct. Look, I have seen many birds who lost their babies return to an empty nest for two or three days and offer food to the empty air. This is not stupidity; she just can't help herself, she has no idea what she is doing or why."

"Are you telling me," I asked, "that it is all simply inherited instinct?"

"Yes," he insisted, "it's just a stereotype, it has no inner significance."

"No feelings attached?" I wondered.

"Zero!" was his response. To drive home his point—which seemed to be how superior we were to ducks—he added, "No human mother would ever behave like this." It is difficult to respond to such an assertion, so generally accepted is this view by science. However, I doubt that Konrad Lorenz himself would agree that there is no subjective component to the behavior. He believed that ducks *felt* constantly, though he would probably maintain that what, exactly, they felt would always remain beyond our understanding.

But one could argue that ducks, observing us, might well conclude that we are such irrational creatures that no possible explanation could attach to our behavior. Indeed, they might feel we are doing what we do because we have been programmed to do so by our genes. Our behavior would have little meaning in duck language to a duck.[154]

After all, ducks and geese have been behaving as they do since long before there were humans interacting with them. William Lishman, who achieved fame by imprinting wild geese to an airplane, then migrating with the flock, writes that "It is deeply meaningful to me that geese, or I should say birds in general, have not gone through any major evolutionary changes biologically for thirty-five to forty million years—a stretch of time which predates by a factor of ten the beginnings of our human species."[155] So it is not surprising that in the short time humans have been around, certain inborn mechanisms of ducks would not have changed. Most of the behavioral patterns of ducks and geese were evolved long before humans were on earth.

The first bird to be domesticated was the greylag goose in India, at some point before 3000 B.C., probably long before chickens and ducks. It has been suggested that geese, and perhaps ducks, were tamed as pets, even as live toys for children. A few centuries later, the mallard was domesticated in the Near East, probably in Mesopotamia. At about the same time, the Muscovy duck in Central America was domesticated. Although the wild mallard does not breed in captivity, its young, if hatched by domesticated birds, will do so. All breeds of the domestic species of duck are descended from the mallard, except for the pied varieties of Muscovy which still closely resemble their wild ancestor. All domestic geese (except for the Chinese goose) are descended from the greylag goose. You cannot get near a wild goose, whereas a tame goose will eat out of your hand and possibly even enjoy your company. According to the *Cambridge Encyclopedia of Ornithology*, this tame quality is "partly due to a somewhat degenerated nervous system caused by an impoverished environment during their development." No artificial environment can possibly be as enriched as a natural one. What domestication for geese and ducks generally means is longer laying seasons, larger clutches of baby birds, and earlier maturity—meaning they mate far earlier than they would in the wild. Tragically, it also often means losing the ability to fly and to walk without difficulty.[156]

In Hungary, several other European countries, and probably (secretiveness is endemic when humans are particularly cruel to animals—think of factory farms, slaughterhouses, and animal experimentation laboratories all of which are notoriously difficult to visit) Israel, there is the live plucking of geese for their feathers. Veterinarians and animal scientists know beyond doubt that this is a cruel practice, and even the experimental literature agrees on the pain involved.[157] Approximately 50 percent of the feathers even on large factory farms come from live plucking. The geese on Hungarian factory farms, where some 20,000 geese are crammed into a small area, are live-plucked (in some cases the birds are hung from their feet) two or three times a year before being slaughtered or force-fed for another year to produce Pâté de foie gras (more on this later).

The collection of down is almost as invasive. All waterfowl have an inner layer of feathers called down. Down is the first plumage on young waterfowl (unlike chickens and turkeys, who do not produce down). It forms a protective insulating liner under the regular feathers of all aquatic birds. It is extremely soft and light and keeps ducks and geese warm in cold water. The insulating property of down comes from what is called "loft"—its tendency to fluff up. Goose down has more loft than duck down, which is why it is used in the production of arctic clothing.[158]

Down clusters are actually not young feathers; down is much lighter than a feather and is three-dimensional, with an unusual construction. The clusters are spherical plumes with hundreds of filaments radiating from a single quill point, like a ripe dandelion pod. Because of its three-dimensional structure and ability to "loft," each down cluster traps air. An ounce of down has two million fluffy filaments that interlock and overlap to form a protective layer of non-conducting air. This makes down a good thermal insulator. This insulating layer helps birds maintain their body temperature in inclement weather. It was clearly designed for these young birds and never meant for human use. Although most down comes from birds butchered for their meat, some are plucked live. The rippers who take these delicate feathers from live birds are none too delicate in their methods. Since they must pluck over one hundred geese a day, you can imagine how careful they are. People sometimes say that since ducks and geese molt and pluck their own feathers (especially the female when preparing a nest for her eggs), there is nothing cruel in plucking their feathers for them. But of course we do not do it for *them* at all, but for *us*. Molting is a gradual procedure, taking several weeks. When a female plucks her own feathers to line her nest, she takes just a few. One author said that these natural processes resemble live plucking as much as yanking all of a child's teeth at once without anesthesia would resemble the natural loss of baby teeth.[159] In other words, not at all.[160]

Eider down is collected in a different way. It comes from the down of the rare eider duck, a large sea duck found in parts of North America, Europe, and Asia. The eider duck has never been domesticated. Eider down is exceptionally soft and has insulating properties superior to any other down. It is taken from Arctic and sub-Arctic eider ducks' nests in the wild, especially in Iceland. The adult birds are not harmed, true, but in order to get the female eider duck to produce more eider from her breast, the "eider farmer" removes both the down and egg. Their ducklings, then, do not hatch. A more ethical way would be to wait until the end of the season and the young have left the nest before collecting the remaining eider. Some Icelandic eider farmers claim to do just this. This would result in much less take, however, and it is the rare collector who is so morally engaged. Nobody, as far as I know, has ever investigated the methods of collecting eider down in Iceland.[161]

Ducks and geese are not only family oriented, they like to be in large congregations (bevies of ducks and gaggles of geese), as do many other species of birds. They have a pronounced gregariousness. This is so concentrated in some species that it has proved possible to photograph the total inhabitants of a continent in a single aerial shot. Such, for example, was a photograph of Pacific black brant showing 174,740 birds and another showing most of the 42,000 greater snow geese still surviving. There is safety in numbers; the likelihood of being selected by a predator is vastly decreased if you are one of several hundred thousand rather than flying on your own. But it also seems the birds are exuberant and enjoy one another's company for the sheer pleasure of it.

Unfortunately, this natural behavior is taken out of context and disingenuously used as proof that intensive duck farming, such as practiced in the United States today, is simply taking advantage of normal duck behavior. The largest duck farm in the U.S.—Grimaud Farms, headquartered in Stockton, California, its parent company in France—operates ten farms raising primarily Muscovy ducks. The wild Muscovy originates in the swamps of South America. They are excellent swimmers, can run very fast, and are able to fly as well at great speeds. At the farms, they are bred to be heavy so that they experience great difficulty walking and are subject to painful leg disorders. Their webbed feet evolved for swimming, but on the farms they are not given access to any pond or even a body of water in which they can dip their whole head. (You can see photos of the farm at www.vivausa.org, the Web site of Viva!USA—a vegan organization that investigates factory farming, or similar photos at www.farmsanctuary.org.) Ducks need water to preen their feathers, and they need to immerse their heads often to clean out their eyes, one reason we see ducks constantly diving even when they are not feeding. Without the ability to preen, for which they need large quantities of water, their bodies soon become encrusted with feces, something never seen in a healthy wild duck. They also develop opthalmia of the eyes, a low-grade infection that can eventually lead to blindness.

A thorough investigation by Viva!USA showed that almost all of these ducks are kept in small cages, have access to only drip water, are forced to eat mush (for which their beaks are not evolved), and are subject to debeaking. Grimaud Farms (and Whole Foods, which carries their products) claim that their beaks are made only of cartilage, and that debeaking is merely like trimming our fingernails. But the Council of Europe—an advisory body of European nations with headquarters in Strasbourg, France, which focuses on such issues as public health, human rights, and scientific matters, including animal welfare—reports that the beaks of Muscovy ducks are richly innervated and well supplied with sensory receptors. Bill trimming, they say bluntly, is mutilation. It is not like trimming nails, more like taking off the tip of your nose, which is also only cartilage, after all. The investigators found an overwhelming smell of ammonia as they entered the area of the cages. The conditions were almost identical to what we have been accustomed to seeing in intensive chicken farming.

Another similar duck farm is Maple Leaf Farms in Colton, California. In 2002, animal rights activists visiting the farm found two little ducklings in a terrible state, covered in dirt and feces, their bills subjected to BHT (bill heat treatment, a way of cutting off the tip of the beak). They took the ducklings with them. Within seconds of access to fresh water, the little ducklings furiously started cleaning themselves and did so for hours after being rescued. By the time they reached their new home at a sanctuary—where they would have a pond to swim in rather than be slaughtered at the age of seven weeks for someone's meal—Jake and Jasper, as they were named, were yellow, as ducklings should be. To this day they remain particularly close to one another.

There is no question, in my mind, that ducks have very good memories. They must in order to migrate. Says H. Albert Hochbaum in his book *Travels & Traditions of Waterfowl*, "When they depart from the lakeshore at dawn, the path of their travel from the very beginning leads them to a particular destination which must have been selected when the journey began." The place lives in their mind, for otherwise how could they make a direct and orderly journey there? "Somehow, quite beyond our understanding, there are within and as a part of each bird all the phases of its life story: the places and companions near and distant both in time and in space, its mother, its first home, its mate, its nest, its home range and territory, its molting place, its wintering area."[162]

Here in New Zealand, we live not far from an enormous natural duck pond at the edge of the sea. It is a protected area, and of course the birds know this and congregate there. The pond is filled with many different kinds of sea birds, but especially with ducks and geese. There is a large New Zealand native bird, the pukeko, with shiny blue wings and a large orange beak, who also comes around in large numbers. I am always amazed to observe how well the different species of birds get on together. Sometimes I will see ten or more species at the pond, but I have yet to witness a single fight. Of course the larger birds will sometimes chase the smaller ones away, but it never comes to more than that. Do they enjoy each other? I would like to think this is the case, although no ornithologist will agree with me. If they were not enjoying themselves, though, why would they crowd so close together, when there is plenty of room for them to be spaced apart? And why would there not be more aggression?

Ducks can form deep and enduring friendships, even across the species barrier, and even when no imprinting is involved. A Muscovy drake developed a strong platonic attachment to a hen named Hetty. When Hetty reached old age and began to fade, the attention of her drake friend was very moving. For the last two days of her life, he never left her side. They remained together in the chicken house, refusing to come out.[163]

Diane Miller from California's Farm Sanctuary told me this story about a friendship between a duck and a chicken: "Ivana is a female Muscovy duck rescued a few years ago. She is an older duck and can no longer lay eggs but still creates a beautiful nest and plops herself down in it every day. Nell is a leghorn hen. Although she is also an older bird, she still lays eggs. Every day during the summer, Nell would walk over from the chicken barn and fly over the fence to the duck and goose barn. She would then work her way into the barn, where Ivana the duck spends 90 percent of her time during laying season—from April through August. Very carefully, she pulls straw around her, forming a nest right next to Ivana and laying her egg (or sometimes not). If she has an egg, and even gets up for a minute, Ivana carefully takes her bill and rolls the egg underneath her. Nell then comes back and sits beside her ducky friend. If Ivana gets up, Nell will just sit on the egg in Ivana's nest, instead of trying to take it away. After staff members saw what was happening, and that on most days neither had an egg, everyone started putting eggs in front of the girls, who carefully maneuvered them into their own nest. Since we do not allow breeding on the farm, we do take all the eggs daily, but make sure to replace any we take from the two girls."

Most ducks, and as far as I know, all geese, are vegetarian. (That is why tame geese are so easy to feed; they mostly eat grass.) So there is little competition for a scarce resource, meat. Aggression is always greater in carnivores than vegetarian species, since hunters need weapons and a willingness to use them. Strangely, though, geese can be more aggressive with humans than are ducks. Ducks, in fact, are not terribly aggressive with other animals, especially if they know them. Darwin in *The Descent of Man*[164] mentions the fact that ducks will lie down and bask in the sun next to a cat and dog they know but run from ones they do not know.

The female duck is normally silent while on the nest during incubation, but as soon as her unhatched chick inside the egg begins to peep, she too makes a quiet squeaking noise. Ducklings and mother ducks respond to each other's calls before emergence from the shell begins. Konrad Lorenz, in *The Year of the Greylag Goose*, heard a goose utter faint "contact calls" to the goslings in their eggs. He believes that goslings are able to produce a number of different calls, which indicate to the mother whether they are developing normally. (Lorenz does not say what happens if they are not, but presumably the mother would roll the egg from the nest.) These same sounds are subsequently used between adult partners, yet another indication that with ducks, it would seem, all love comes from mother love.

Young goslings and ducklings produce a variety of sounds, indicating distress call or comfort. A single shrill whistle is given when they see aerial predators. This means both "Help!" and "Watch out!" Indeed, as soon as the sound is produced, other young birds in the nest crouch to make themselves less visible and to be out of the way of a grasping hawk claw. Whether this sound is learned or innate is not clear. Lorenz was so impressed with the first of these calls, that he entitled his last book summing up his work with geese *Hier bin Ich, wo bist Du?* (*Here I am! Where are you?*) Jean Delacour, the great French ornithologist, believes we should add "conversational notes" to the repertoire of geese.

W. H. Hudson, one of the greatest writers on birds, was, like many of us, fascinated with the fidelity of the goose. He tells a particularly touching story he heard from a man who was sheep farming in a "wild and lonely district on the southern frontier of Buenos Aires." One day this man was out riding, just after all the flocks had flown south, when he saw at a distance before him on the plain a pair of geese. "The female was walking steadily on in a southerly direction, while the male, greatly excited, and calling loudly from time to time, walked at a distance ahead and constantly turned back to see and call to his mate." Then the gander would fly into the air and turn back to see if his mate was following him. Alas, she could not; her wing was broken. She had set out on the long journey to the Falkland Islands by foot. He would not leave her, so after flying for a few hundreds yards, he would alight and wait for her to catch up. He would fly ahead, to show her the way, then return "again and again, calling to her with his wildest and most piercing cries, urging her still to spread her wings and fly with him to their distant home." He knew their fate: "In that sad, anxious way they would journey on to the inevitable end, when a pair or family of carrion eagles would spy them from a great distance—the two travelers left far behind by their fellows, one flying, the other walking; and the first would be left to continue the journey alone."

Geese are so wary of danger that "enemies" rarely get close. When feeding, at least one goose is always on the watch and warns the feeding flock of danger. Nobody I asked seemed to know whether it is a gander or the goose who volunteers (if that is what happens) for this task, but I suspect it is the gander. Geese have even stronger family bonds than swans. They tend to remain paired for life (which is long; a captive Canada goose lived to over thirty-two years, but there are scientists who believe the marriage between wild geese can last for nearly a century). Geese fifty years old have been known to breed with one another.[165] Do geese mourn the loss of a partner? Of course. Konrad Lorenz had no doubt of this, nor should we.

When the goose is on the nest, the gander will float offshore, acting as a sentinel. It is almost impossible to surprise a goose. They have remarkably acute powers of sight and hearing. This is why so many farmers keep them: they are wonderful watchdogs (they have been so used for many centuries), honking loudly when anyone unfamiliar approaches the farm. Unlike the dog, however, they are not warning a human "master" but other members of the goose flock. In the wild, whenever geese are congregated somewhere, other birds like to come as well, no doubt because they benefit from this early warning system.

The sound of the goose is not just a warning. For many humans it is poignant in ways they have trouble explaining. W. H. Hudson recalls how he became fascinated with geese when he was a small boy.

Not far from where he lived in the country in England was an old mud-built house, thatched with rushes, at the edge of a marsh that was a paradise of wild fowl including swans, half a dozen species of ducks, plovers, godwits, and the great crested blue ibis. Three ancient women, a mother and her two old daughters, inhabited the house. According to Hudson they looked and acted like witches, except when they were among their beloved birds. Most loved of all were the wild geese, who had become completely tame because of the gentle way these three isolated women treated them. Hudson was fascinated from the beginning, both by the geese and by their relationship with these three women—whom other humans avoided, but who found their true love in wild geese. Later, he was to recall how a distinguished man of letters was once asked why he lived away from society, buried in the loneliest village on the dreary east coast of England, on a spot looking over a flat desolate shore to the North Sea. He answered that he made his home there "because it was the only spot in England in which, sitting in his own room, he could listen to the cry of the pink-footed goose." Hudson responded, "Only those who have lost their souls will fail to understand."[166] Hudson lived for many years in Patagonia, in the south of Argentina (where his famous *Green Mansions* was set). He would often, on a still frosty night, listen to the birds (Magellanic geese) all night long as they flew low, following the course of a river. "To listen to their wild cries, I would willingly give up, in exchange, all the invitations to dine which I shall receive, all the novels I shall read, all the plays I shall witness, in the next three years."

There is another sound of great significance to geese (not, as far as I know, for ducks), that Lorenz called the sound of the triumph ceremony. This is heard whenever a gander drives off an intruder. At such a time, he invariably utters a triumph note and equally invariably the female repeats the note and stretches out her head and neck close to the ground. The young, including even down-covered babies, show their appreciation by performing in the same approving way. "Look what I did!" is the unmistakable meaning of this ceremony. A version of this triumph ceremony can be seen when two geese mate. The gander swims in a particularly proud way, his wings slightly raised, his tail cocked, and his neck curved. He dips, a kind of bow, to the female. Then they raise their wings together, extend their heads straight into the air, and give a honk of joy or celebration or achievement to announce that the deed has been done. The gander does not help with the incubation of the eggs, but he does stand guard at the nest and helps raise the goslings, much as swans do.

The expression "silly goose" does not mean what it is generally taken to mean. Yes, today it means a person who is foolish or weak-minded, but this is a false reading of its original sense. The *Oxford English Dictionary* says the original meaning of "silly" was deserving of pity, compassion, or sympathy. So, the silly goose was one about to be killed, for whom we should feel compassion. I plead for a renewal of this wonderful original version!

I made the mistake of asking a French farmer what he loved about geese and the answer I got back was "their liver." Pâté de foie gras, fattened goose liver, is obtained in a horrible manner.[167] We must not make the mistake of thinking this is merely a modern practice, and that in earlier times people, especially farmers, were kinder to their geese. This mistaken notion was put to rest with the appearance of the classic book by Keith Thomas, *Man and the Natural World: A History of the Modern Sensibility.* He writes, "Battery farming, moreover, is not a twentieth-century invention. . . . Poultry and game-birds were often fattened in darkness and confinement, sometimes being blinded as well. . . . Geese were thought to put on weight if the webs of their feet were nailed to the floor; and it was the custom of some seventeenth-century housewives to the cut legs off living fowl in the belief that it made their flesh more tender."[168]

It is not an innovation of nineteenth-century Europe either; Pliny, the Roman naturalist of the first century A.D., recorded that "our people esteem the goose chiefly on account of the excellence of the liver which attains a very large size when the bird is crammed." In the second century A.D., Plutarch the Greek recorded how the eyes of swans and cranes were sewn together to "shut them up in darkness." An account of the nineteenth-century treatment of Rouen ducks in Brittany and Normandy tells how the "poor birds are *nailed* by the feet to a board close to a fire and, in that position, plentifully supplied with food and water. In a few days, the carcass is reduced to a mere shadow, while the liver has grown monstrously."

Today, in France, Hungary, and Israel (as well as in the United States), a mixture of maize, fat, and salt is force-pumped down the gullets of geese and (increasingly) ducks.[169] In most places in France, this force-feeding is mechanized, with machines doing this with a metal tube for at least sixteen days (in some cases, twenty-eight days)—three times a day until the twelfth day, then every three hours even through the night. About 30 percent of birds kept for foie gras are force-fed mechanically. The vast majority of ducks bred for their livers are fed pneumatically. On some farms the ducks or geese are kept in near darkness, presumably to keep them disoriented. Pneumatic force-feeding systems can cram up to a pound of feed down a duck's throat in three seconds.

Eighty percent of all ducks raised for foie gras production are kept in individual cages in which they cannot turn around, stand up, or stretch their wings properly, notes a major study produced by Advocates for Animals and the World Society for the Protection of Animals, a consortium of more than 300 member organizations in 70 countries to promote animal welfare.[170] The grossly engorged and diseased liver swells up to ten times its normal size. Force-feeding causes a fatty degeneration in the cells of the liver. These birds are suffering from a serious liver disease. They have difficulty standing and can barely walk. It would be like force-feeding thirty pounds of spaghetti a day to a human being. Needless to say, many times the stomachs burst and the birds die, or their throats are perforated by the pipe. (When the feeding is done by hand, workers are expected to force-feed 500 birds a day.) One American veterinarian who observed the procedure wrote to the WSPA: "Foie gras is produced at a terrible cost to the birds themselves. Foie gras, touted as a gourmet delicacy to entice the palate, is really only the diseased tissue of a tortured sick animal." Dr. Mark Lerman, another U.S. veterinarian, wrote: "The lesions seen in this duck and others like him, are unique. They are the result of a continuous, perverse and concerted effort to physically force these poor creatures to do something they weren't designed to do."[171]

More than twenty-five million ducks and geese annually are subjected to this torture. Surprisingly, even though more and more people are learning just what is involved in foie gras production, since 1990 the French foie gras industry has doubled! France exports to nearly 100 countries worldwide. In the Loire region, production soared from 121 tons in 1990 to 2,032 tons in 1998. The European Union's Scientific Committee on Animal Health has condemned the practice, and the force-feeding of birds has been prohibited in many countries in Europe, including Denmark, Norway, Austria, Switzerland, and Poland.

I tried to discuss this cruel practice with the same French farmer who told me he liked goose liver. He insisted that it was not cruel, that it was simply taking advantage of the natural inclination of geese. Ducks and geese are prone to diseases of the liver because as water birds they naturally carry more fat beneath the abdominal skin than other birds. Since they are migratory birds, they have a tendency to eat more and lay down stores of lipids in the autumn and spring. But of course the birds never eat enough to harm their own livers; at the most the liver doubles in size and soon resumes its normal size. French culinary authors like to say that "it was the goose itself that invented cramming."[172] Sonoma Foie Gras proprietor Guillermo Gonzalez claims that while caviar and truffles are products of nature, "foie gras can only be produced by enhancing the duck's or goose's ability to develop a large, firm liver." True, the farmer conceded to me, the day after the force-feeding is finished the birds are slaughtered, but they would be slaughtered in any event. "Could I see how it is done?" I asked, but he refused. In the end I had to be satisfied with watching a video. It was more than one can bear.

In his book on Henrik Ibsen[173], Michael Meyer points out that the symbol of the duck in the play *The Wild Duck* came to Ibsen after he read in Darwin that ducks degenerate in captivity. Presumably Ibsen had in mind the passage in *The Variation of Plants and Animals Under Domestication*, where Darwin speaks of an excellent observer "who has often reared ducks from the eggs of the wild birds found that he could not breed these wild ducks true for more than five or six generations, as they then proved so much less beautiful; they increased also in size of body; their legs became less fine, and they lost their elegant carriage."[174] The word "degeneration" does not occur often in Darwin. Clearly he believed here that domesticated forms were degenerated, whether, as he says elsewhere, because of lack of exercise, poor diet, or other causes. The most serious consequence, however, is the loss of flight.

Evidently it did not occur to Darwin to wonder how these losses affect the minds of ducks and geese, something left to our generation to mull over. Ibsen himself added an image of remarkable poignancy: When ducks are shot, he says, "They always dive to the bottom as deep as they can get, and tangle themselves in the muck and seaweed, hanging on with their beaks to whatever they can find down there. And they never come up again." What is it that domesticated ducks and geese feel? Well, they feel everything that their wild ancestors feel, except that they do not have the opportunity to take the actions that prompt the feelings.

Ducks and geese are aquatic birds; their whole life revolves around large bodies of water. Imagine being deprived of the joy of swimming, diving, racing around a pond or a lake. They are also, in their wild state, powerful fliers. The ability to fly has been bred out of most domestic ducks and geese, but this does not affect the desire to fly and to feel the accompanying emotions. Imagine how it feels to see your wild cousins flying overhead and to realize that you cannot join them, even though you feel the same wild urge to leave. Migration is so deeply built into the genes of most ducks and geese, that it must prove unbearable not to be able to exercise this instinct. Remember the old song: "She'll see a shadow fly overhead, she'll find a feather beside my bed." This itch to be going, this wanderlust, has a separate word in German—*Zugunruhe*—used only of birds that migrate but no longer have the ability to do so. It was first described by Johann Andreas Naumann at the beginning of the nineteenth century. He interpreted it as an expression of the migratory instinct.

Indeed, the urge to migrate is at least partly coded in the genes. "Caged birds," as William Fiennes notes in his popular book *The Snow Geese*, "get restless in a particular direction." Caged birds, he notes "seem to act out the entire course of the migration [of their wild cousins] in the confinement of their cages." We can imagine that this inability produces a profound sadness. To be deprived of your very birthright is to feel a profound absence. Like a phantom limb, the wings feel the stirrings and then . . . nothing. Domesticated mallards can often be seen racing to a pond; suddenly they are able to take off, but just barely. Madly flapping their wings, they manage to coast a few hundred feet just above the ground. It must be exasperating for them. Do they have an atavistic memory? I would not rule the idea out. So basic to their nature is flying away (and returning) that to be deprived of the ability to do so must leave every domestic duck and goose bewildered, frustrated, and depressed.

How can we reconcile the barbaric methods used to obtain foie gras and down feathers with the inspirational sight of wild geese flying in their famous V-shaped formations or trailing lines, led by an old bird? Geese in this respect are like elephants; the oldest knows the best migration route from experience. Sometimes, too, geese switch the lead positions regularly to conserve energy, like bike riders drafting.

Climbers on Mount Everest are reportedly struck with awe as they approach the summit of the mountain only to see wild geese silently passing overhead, some six miles above sea level. In fact, geese fly higher than any other bird, except swans. This was conclusively demonstrated in 1970 when a flock of bar-headed geese were recorded flying over the Himalayas. The flock migrated directly over the summit of Makalu (the fifth highest mountain in the world, just a few miles from Everest), 27,824 feet above sea level. (By contrast, humans are in a state of beginning hypoxic collapse at 20,000 feet and will fall into a coma and die at 23,000 feet; birds, however, have a remarkable breathing system with extra air sacs and air space in their hollow bones.) Humans react with something more than our minds when we observe a flock of geese cruising at extraordinary speeds high in the sky. But while these regal birds may inspire us for a brief moment, to our shame few of us honor them in our daily lives.

SIX

The Nature
of
Happiness

In order to discover what is best for an animal, we are to a great extent dependent upon our ability to imagine ourselves in the life of that animal. Just about everyone would agree that it is better for an animal to be happy rather than unhappy, regardless of what our position is about other matters concerning animals. The difficulty comes when we attempt to define happiness. Philosophers and animal welfare scientists have generally approached this task with skepticism, saying that we cannot really know what makes any given individual animal happy, that the word is notoriously difficult to define, that we could not even give a definition of human happiness that would satisfy everybody, and so on. This is disingenuous; we know there are certain things that nobody is happy about.

In an essay on death, the philosopher Thomas Nagel writes, "Life can be wonderful, but even if it isn't death is usually much worse. If it cuts off the possibility of more future goods than future evils for the victim, it is a loss no matter how long he has lived when it happens. And in truth, as Richard Wollheim says, death is a misfortune even when life is no longer worth living." Few people would quarrel with these thoughts. But when it comes to animals, we hesitate to accord them the same weight.

Some philosophers claim that if an animal cannot have an expectation about its future, it is somehow less deserving of our concern. Since a position similar to this has been ascribed to the philosopher Peter Singer, a leading figure in the revival of the animal rights movement, I asked him whether he really believed it. He wrote to me that "If we are considering whether a being has a right to life—or as I would prefer to put it, if we are asking how great a wrong it is painlessly to take the life of a being—we need to ask how great an interest it has in continuing to live. Here I think it does make a great difference if a being can see itself as existing over time, and have a sense of its own future. Only then can it have preferences to continue to live (as distinct from preference to avoid certain threatening or painful situations) and only then can killing it render nugatory or pointless much of its previous activity." But an animal who lives in the present and has few expectations—a dog, for example, though dogs certainly expect to be taken out for a walk—is no less important or worthy of our concern and love any more than is a person who thinks primarily about the here and now rather than what will come later.

Other philosophers continue to maintain this point of view: that since an animal cannot conceive of its death, it cannot suffer injustice when it is subject to it. But Nagel points out, justly I believe, that when it comes to death for a human, there is nothing to expect. How, he asks, can we expect nothing *as such*? We do not. We all have great difficulty in imagining a time when our present existence will simply cease. It is, in some deep sense, unimaginable; we avoid the thought. The fact that an animal does not avoid it, but simply cannot begin to think about it, is really neither here nor there when it comes to the ethical questions of whether life would be better prolonged than cut short. For us, as for them, life is almost always to be preferred over death. And how can the previous activity of an animal be rendered pointless in any event? How many of us would have a convincing answer were we asked, "What has been the point of your life?" If we feel, in middle age, that we have wasted our life, or if we agree with Rilke that "you must change your life," does that mean that what preceded was pointless? That seems rather harsh either for an animal or a human.

Many animal behaviorists and other biologists consider the question of animal happiness to be pointless. We can never know, they say, what it takes to make an animal happy. I think we can know quite easily. An animal is happy if he or she can live in conformity to his or her own nature, using to the maximum those natural traits in a natural setting. To live according to nature will differ for each species, but the answers are not unfathomable mysteries. A chicken likes to take a sunbath. She will roll onto her side, extend one wing, exposing it to the sun, then roll over onto her other side and do the same. This clearly gives her enormous pleasure. It is natural behavior. A chicken evolved to do this. A chicken who never sees the sun cannot be said to be happy because she cannot engage in behavior she was meant to perform. A chicken spends hours scratching in the dirt; a rooster protects young; a hen raises her chicks. All these conditions make for happiness even if that experience cannot be expressed or even known. Nobody can be happy who does not live according to the dictates of his or her nature. (For humans, of course, what is natural is much harder to decide, perhaps because we have choices usually not available to other animals, e.g., birth control. But consider how we would feel if basic human rights were abrogated, if someone else decided when or if we could become pregnant or if women were forced to bear children for the state or an employer.) Most birds were born to fly. (If it makes you happier to be more scientific, you may say the bird evolved to fly.) A bird in a cage is not happy because that bird is not flying, something the bird was meant to do. Even if birds experience moments of happiness, because they cannot fulfill their true nature we cannot say they are leading a happy life. This is why living with a parrot is an ethical dilemma, but not so living with a dog. Chickens evolved to perch in trees at night to avoid predation and to have the warmth and comfort of friends. They are not happy when confined six to a cage. Cows are herd animals. When transported in dark trains to the slaughterhouse, they cannot help but feel panic.

The question of animal happiness is hardly a trivial matter from a philosophical point of view, a scientific one, or a moral one. Gone are the days when we could quote Bertrand Russell, who began his book *The Conquest of Happiness*, published in 1930, with the patronizing comment that "Animals are happy so long as they have health and enough to eat." However, recent comments by the philosopher Roger Scruton that farm animals who are housed together in the winter and allowed to roam in the summer "are as happy as their nature allows," are not so different.[175] How can Scruton, or anyone else, arbitrate what limits there are to the nature of any other creature? More subtle versions of this viewpoint persist. This is odd, because the first generation of English ethologists, such as W. Thorpe of Cambridge University, had considerable influence in urging the British government to permit all domesticated animals to express their "natural instinctive urges." When a high-level committee known as the Brambell Committee was formed decades ago to advise the British government on how much suffering animals experienced when they were confined, the answer was not difficult. Suffering was easy to assess. Sensibly, the scientists agreed than an animal's suffering corresponded to "the degree to which the behavioural urges of the animal are frustrated under the particular conditions of confinement." Unfortunately, more recently there has been a shift in part of the scientific world as if suffering were a completely mysterious feeling, one about which we can know almost nothing.

Many animal welfare experts do not agree with the point of view that the more natural the situation of an animal the happier that animal will be. Ian Duncan and David Fraser have mounted a powerful criticism when they write:

> The concept of an animal's "nature" would need to be made more specific before it could give clear guidance in judging animal welfare; generalizations might lead us astray. For example, we might conclude that seagulls (*Larus*) had evolved to live in such close association with the sea that this is an essential part of their "nature." However, within the past 30 years, the herring gull (*Larus argentatus*) and the less black-backed gull (*Larus fuscus*) have changed their habits in northwestern Europe and now voluntarily live in very artificial environments created by human beings: they nest on buildings, roost on playing fields, and forage on garbage dumps. Gulls are so successful in this mode of living that such populations are expanding rapidly. Thus, it turns out that the sea is not an essential part of the "nature" of these seagulls.[176]

But the life these seagulls live is essentially the same, even if the habitat is slightly different. They are hardly in captivity even if they are not living on the ocean. The artificial environments are still natural in that the animals themselves are not in confinement. They are free and freedom is an essential ingredient of life in nature.

There is a new direction in writing about animals: many writers take it for granted that animals have consciousness and feelings, and can suffer. The change began with Darwin's 1872 masterpiece *The Expression of the Emotions in Man and Animals*, but for reasons not entirely clear that book and that direction fell into obscurity. Scientists such as Donald Griffin, Jane Goodall, and Frans de Waal have revived it. At the same time a number of naturalists were writing books about particular animals that were decidedly different from those of their colleagues. Books by Hope Ryden about the beaver, *Lily Pond*, and by Joe Hutto about the wild turkey, *Illumination in the Flatwoods*, are outstanding examples of a new breed of animal observation: participatory animal anthropology, it could be called.

This kind of writing has been the subject of much criticism. Some hard-nosed scientists even deny that animals are conscious, though this position is losing ground daily. Many still consider the subjective life of animals beyond human ability to research. Needless to say, this same criticism may be directed at a book about the emotional lives of farm animals. Virtually nothing at all has been written about the feelings of farm animals; the reason must be that certain attitudes are deeply entrenched. This is very odd. However, I suspect in the next few years, we will see more books on this topic. There are trends of fashion in science like anywhere else, and it seems like scientific folly to ignore this topic.

There was a time, not so very long ago, when people intimately connected to the lives of animals did not care whether animals had feelings or not. Frans de Waal, from the Yerkes Primate Center at

Emory University, writes in a 1999 editorial in the *New York Times*, "I still remember some surrealistic debates among scientists in the 1970s that dismissed animal suffering as a bleeding-heart issue. Amid stern warnings against anthropomorphism, the then-prevailing view was that animals were mere robots, devoid of feelings, thoughts or emotions."

Jaak Panksepp, one of America's leading neuroscientists, strongly believes in the existence of emotions in animals. He begins his landmark book *Affective Neuroscience: The Foundations of Human and Animal Emotions*, by observing that there is no doubt that both animal and human brains are wired for dreaming, anticipation, the pleasures of eating, anger, fear, love and lust, maternal acceptance, grief, play and joy, and "even those that represent 'the self' as a coherent entity within the brain."[177]

All of us have felt deep emotions that we cannot put into words. These often depend upon memories of physical sensations—for example, lying in the sun, sitting by the seashore, hearing the voices of children, seeing the leaves fall in autumn, listening to birds singing in the morning. These are all triggers we share in common with other animals, and it is not a big jump to assume that animals have equally powerful memory and feelings. We can also feel strong emotions triggered by memories of being with someone, eating together, making love, sleeping together, taking walks together. Again, we share these experiences with other animals who also have friendships. At the deepest level of our being, we are remarkably like other animals, for these feelings are not thought-mediated. This allows us to bypass altogether the debate about whether other animals think as we do. Whether they do or not is irrelevant to the question of whether they feel as we do.

The human urge to compare ourselves as a species with other species is old but is not shared, it would seem, by any other species. Darwin, in a passage in *The Descent of Man and Selection in Relation to Sex*, says that the highest animal would have to admit that what sets humans apart from other animals is that these other animals could not "follow out a train of metaphysical reasoning, or solve a mathematical problem, or reflect on God, or admire a grand natural scene." However, contradicting himself on this last point, Darwin admits that apes can admire the beauty of the fur "of their partners in marriage." (It seems no one has noticed that Darwin, of all people, refers to animal marriage![178]) But let's look more closely at Darwin's list of human characteristics. Most of us cannot follow metaphysical reasoning, or don't want to. As for reflecting on God, many humans—in particular the very ones Darwin would most have admired (scientists)—would consider this a complete waste of time. That leaves mathematics, no doubt a considerable human achievement, but it is nonetheless interesting that of the four characteristics supposedly uniquely human, we are left with only one. We ought not to forget that it was Darwin who was willing to make the most profound comparisons between humans and what were then still called "the lower animals." Later he was to pencil a note to himself: "Never use the word [sic] higher and lower."[179] Good advice to this day.

True, humans have the capacity for metaphysical reasoning, but it is not at all clear to some people that this is such a great advantage that it should single us out from all other animals. From an evolutionary point of view, metaphysical reasoning does seem something of a dead end, which may well be why we do not find it in cows, or at least we don't think that is what they are thinking about when they seem to be lost in meditation. Darwin himself, one of our great intellectual heroes, could not fathom his own feelings in at least one significant instance. His mother, Susanna, died in 1817 when he was eight years old. Nobody in the family ever mentioned the mother again. Years later, Darwin wrote to a cousin who had just lost his young wife: "Never in my life having lost one near relation, I daresay I cannot imagine how severe grief such as yours must be." Did Darwin forget his mother, or were his emotions completely unavailable to him at the time and later? Intelligence, then, is no guarantee that emotions can be assimilated, faced, or even felt.

Perhaps the best comment I have seen on anthropomorphism comes from C.W. Hume, the founder of the University Federation of Animal Welfare in England, who said:

> To this day the more conventional biologists suffer from an obsessional fear of anthropomorphism, and even put such words as "hunger" and "fear" between quotes (a literary solecism in any case) when writing about animals. The quotes are a way of saying "I cannot get on without Anthropomorphism, but I am ashamed to be seen with her in public."[180]

The attitudes toward animals that we have inherited until only very recently are a result of behaviorism. Behaviorism, the dominant paradigm in the United States for nearly fifty years, takes the position that it is scientific nonsense to seek out subjective experiences. All that matters is observable behavior. Behavior, it is thought, can be studied with supposed objectivity, it can be controlled, and it can be replicated in a laboratory.

John B. Watson, the founder of behaviorism, wrote, in the very first line of his book *Behavior*, that "Psychology as the behaviorist views it is a purely experimental branch of natural science. Its theoretical goal is the prediction and control of behavior. Introspection forms no essential part of its methods . . . " He is more explicit a few pages later: "It is possible to write a psychology . . . and never to use the terms consciousness, mental states, mind, content, will, imagery, and the like."[181] Questions about what animals might be thinking and feeling became the very type of question that was not permitted to be asked. Watson's successor, and the founder of what is called "radical behaviorism" (internal states are not permitted), was B.F. Skinner, who had enormous influence on American psychology. When he was interviewed by Bernard Baars, the American neuroscientist, Skinner stated: " . . . it is possible . . . to be a behaviorist and recognize the existence of conscious events. . . . But I preferred the position of *radical behaviorism*, in which the existence of subjective entities is denied."[182]

In a recent book about the remarkable abilities of an African grey parrot, Alex, Irene Pepperberg notes, "In the 1970s, the behaviorist tradition, best exemplified by Skinner, still represented mainstream, laboratory-based behavioral science. . . . According to Skinner, one needn't study a wide variety of animals, because none would react any differently from a pigeon or a rat: The rules of learning were universal."[183]

When ethology, the scientific study of animal behavior, came of age in the last years of the Second World War, those who belonged to the new discipline had to confront this doctrine, and in print, at least, Niko Tinbergen complied, as a famous stern passage from his 1951 book *The Study of Instinct* reminds us: "Because subjective phenomena cannot be observed objectively in animals, it is idle either to claim or to deny their existence."[184] In some circles, it is still contentious to say of a cockatoo that she really enjoyed being scratched behind the ear!

This behavioralist era is now officially coming to a close. The distinguished neuroscientist Antonio Damasio, in a recent highly praised book, *The Feeling of What Happens*, writes that "The conscious mind and its constituent properties are real entities not illusions, and they must be investigated as the personal, private, subjective experiences that they are."[185] I asked him what he thought of the possible superiority of some animal emotions, and although he found the notion intriguing, he told me, "I am inclined to believe that animals do not have 'superior' emotions. It is certainly the case however, that emotions play a greater role in animals than they do in humans, given their immense power, and given the fact that animals cannot rely on our prodigious capacities of memory and language."

Given the long reign of behaviorism, it is not surprising that almost nothing has been written on the subjective lives of farm animals, despite their having been with us so long, and on such intimate terms of life and death. But then, think of how much was known in the past about the inner lives of slaves. It is not only fashion that changes, but our capacity to assimilate knowledge. After all, first-hand accounts of incest are not to be found at all in the nineteenth century but are abundant today. Behind the lack of accounts was a scientific establishment here as well: psychiatry. The psychiatrists tried hard to keep these realities from shattering what were merely prejudices they had been taught as scientific truths. New and revolutionary knowledge almost always has that effect on inflexible categories, especially on the borderland between science and everyday life. Sometimes it is hard to see what science has to do with any of this: slavery, the hurt of women, physical abuse of children, and animal lives.

The charge of sentimentalism is one often brought against people interested in animal issues and was probably first raised in the nineteenth century, during debates about vivisection (animal experimentation that caused animals great distress). In the 1970s and 1980s, Juliet Clutton-Brock was also often accused by her colleagues at the Museum of Natural History of being sentimental. She observed that nobody thought the men who worked in abattoirs were sentimentalists. So do you have to be willing to kill animals without feeling to be considered free of sentimentalism? I brought up the question with the British philosopher Mary Midgley who pointed out to me that if we are angry with someone whom we see beating a child we do not call that emotion unsuitable. Well, would we call that emotion not suitable if we saw someone beating a dog? Probably not. In fact, she mused, to be angry is almost never called sentimental. We reserve this insult for the "positive" emotions. It is obviously not self-indulgent to be upset at seeing someone abuse a child. We are permitted our sentiments when it comes to beings who are not rational adults. But if we were to say that we felt compassion upon seeing a pig being led to slaughter, we would immediately be accused of sentimentalism.[186] Darwin was in fact very angry when he saw someone beating a horse, as any proper Englishman would be. That would not be called sentimentalism but appropriate rage at something unacceptable. It all depends on our interpretation of whether the object of our sentiment is within the circle of our concern. If so, then the feelings we have are appropriate; if not, they are not.

This is not very logical. Part of the problem, as Mary Midgley has pointed out, stems from the fact that sociobiologists have insisted for some time that "we are born selfish," which makes any kind of sentiment associated with the positive emotions appear unrealistic. By definition, sentimentalism can involve indulging in superficial emotions. Nobody wants to defend this. On the other hand, the word has been used from time immemorial to describe something negative. Kenneth Clark, in his popular book *Animals and Men*, begins the chapter on "Animals Beloved" with this comment: "The love of animals is often spoken of by intellectuals as an example of modern sentimentality." He counters: "I may therefore begin this section by quoting the feelings of a man the greatness of whose mind is not in doubt, Leonardo da Vinci. 'He greatly delighted in horses, and in all other animals, which he controlled with the greatest love and patience, and this he showed when often passing by places where birds were sold, he would take them out of their cages, and having paid those who were selling them the price asked, would let them fly into the air, giving them back their lost liberty.' This well-known passage in Vasari is confirmed in Leonardo's Notebooks, and particularly by a reference in a book by an early traveler named Corsali who, writing in Leonardo's lifetime of a gentle savage tribe, says that they will not eat the flesh of any living thing, like our Leonardo da Vinci. So Leonardo not only loved animals, but also, with a consistency rather rare among animal lovers today, was a vegetarian."[187]

The charge of sentimentalism often arises with the use of the word "love."[188] This word clearly makes some scientists nervous. But it did not do so for Darwin. In *The Descent of Man*, Darwin says explicitly: "Most of the more complex emotions are common to the higher animals and ourselves . . . Animals not only love, but have the desire to be loved." It is in this famous Chapter Three, about the moral sense, that Darwin comes closest to the modern attitudes of those who wish to give animals deeper significance in human moral development, when he says that "Sympathy beyond the confines of man, that is humanity to the lower animals, seems to be one of the latest moral acquisitions. . . . This virtue, one of the noblest with which man is endowed, seems to arise incidentally from our sympathies becoming more tender and more widely diffused, until they are extended to all sentient beings." Darwin could almost be considered a Buddhist here. How astonishing that he should use the term "sentient beings" for animals, when that is precisely the position that the European Union has finally taken, after much prodding, especially by the advocacy group Compassion in World Farming in England.[189] No doubt Darwin was led to this position by his strong belief that animals share the same emotions that humans do. In his notebooks he suggests the mind and emotions must be the same for animals and humans, the very essence of my own thesis. Darwin is most explicit in his comments on the mental powers of animals when he writes:

> It has, I think, now been shewn that man and the higher animals, especially the Primates, have some few instincts in common. All have the same senses, intuitions, and sensations—similar passions, affections, and emotions, even the more complex ones, such as jealousy, suspicion, emulation, gratitude, and magnanimity; they practice deceit and are

revengeful; they are sometimes susceptible to ridicule, and even have a sense of humor; they feel wonder and curiosity; they possess the same faculties of imitation, attention deliberation, choice, memory, imagination, the association of ideas, and reason . . . "

It was taken for granted, in the period immediately preceding and following Darwin, that animals had emotions. If critics are going to say that people like myself who attribute complex emotions to animals are guilty of anthropomorphism or sentimentalism, they will have to apply the same criticism to Darwin himself. It may have originated with Darwin's young friend George John Romanes, who from 1874 until Darwin's death in 1882 was his research associate.

Just as bad, according to some scientists, is the use of anecdotal evidence. When, exactly, did this become a sin? We can point to a passage in Romanes's 1882 book, *Animal Intelligence*, where he repeats an account of ant funerals that claimed that ants carry their dead in a marching procession, dig graves for them, and then individually bury them! Romanes claimed this was an observation "about which there could scarcely have been a mistake." A contemporary, Margaret Floy Washburn, professor of philosophy at Vassar College, in her book *The Animal Mind* commented, dryly: "One is inclined to think it just possible that there was."[190]

The fact that Romanes misused anecdotes does not necessarily make them unusable. I would like to attempt a resurrection of anecdotalism. True, the word itself implies triviality: "an amusing little anecdote." Years ago I would have been accused of being anecdotic, or addicted to anecdotes. Is there a progression from an anecdote to an account, to an observation, to a case study, to an ethogram, the precise, detailed description of every movement and behavior of an animal? A field biologist, by definition, is someone who goes out into the field to observe animals. The problem is that the method he or she uses, the ethogram, is boring. A case study is equally so, and in fact one begins to think that to be scientific one must be boring. Perhaps we need to reclaim the narrative power of Darwin—casting his net wide, observing, asking, tying disparate elements together, and searching for the bigger picture in the smallest details.

However, we are not likely to hear much about the extraordinary abilities of animals whose purpose in the eyes of the teller is merely to be served as lunch. The cognitive dissonance, as psychologists call it, would be simply too great. And indeed, my experience has been that asking for colorful anecdotes about farm animals who are about to be slaughtered is not productive.

Freud used the explanation that people who could not accept the reality of the unconscious were engaging in a kind of denial. He called it resistance, i.e., resisting the truth. His critics jumped on him, because of course it is very easy to dismiss people who do not believe you by saying they are resisting the truth. But he had a good point. People who deny the reality of animal emotions seem to me to be engaging in resistance, in this case resisting the fact that we are forced, whether we like it or not, to use our own sympathy, empathy, and imagination to attempt to put ourselves in the place of another person or animal. While anecdotalism and anthropomorphism and sentimentalism are worth thinking about, still we often are dependent upon these very human failures in order to imagine our way into the existence of another life form. We can do it badly or we can do it well, we can do it with knowledge or with ignorance, but we cannot entirely escape it, however hard we try.

CONCLUSION:

On Not Eating Friends

What I have tried to do in this book is to show that cows, chickens, goats, sheep, ducks, geese, and pigs (and by implications all other farm animals) have the same capacity for emotional complexity and intensity as did their evolutionary ancestors. If these emotions are not always visible to us that may be because we are simply not capable of seeing them, or because we have placed these animals in situations where they cannot express the emotions they inherently possess.

I have used a simple device throughout this book: In order to understand an animal we know today, I have looked at its ancestors of yesterday. I do this because when I was writing about dogs it soon became obvious that in order to understand the inner life of dogs I had to know something, in fact as much as possible, about the lives of wolves, because dogs are descended from wolves and their behavior is basically the same. Understanding one leads to a deeper understanding of the other. This seemed to me especially true when the animal you wish to study does not make his or her inner life apparent, as is the case with farm animals. They appear to us as remote, inaccessible—partly because we have never made the effort to get to know them on their terms, and partly because we do not live with them and around them the way we do with dogs and cats. We do not study them; we kill them. What does it take to become everybody's favorite animal? You need to become acquainted up close. The more we learn about farm animals, the more profound they appear to be. The less we know, the less important their lives. It is impossible to know them and not to respect them, not to care about how their lives go for them. It is impossible to realize that their ancestors were capable of deep and complex emotions, such as nostalgia, compassion, love, joy, disappointment, and others, and not wonder whether the lives they live permit them to express such emotions.

What has any of this got to do with you, you might ask? If you eat these animals, if you wear their skins as shoes or belts, then their lives must be of concern to you. It has something to do with you, because you have something to do with them. Our lives, all of our lives, are inextricably intertwined with the lives of farm animals, even when we would prefer that they not be. It would take a very hard-hearted person to say: "I don't care what kind of lives they lead, how much they suffer, how far removed from their ordinary life, it just means nothing to me, holds no interest for me. I will continue to eat them and use them in any way I feel like without taking the slightest responsibility to know what kind of creatures they are, what they feel, what kinds of lives they lead in order to give me the products I want." Very few people would ever say such a thing. In fact, a recent Zogby poll shows clearly that the vast majority of Americans care very much about how the animals they eat live.[191]

Why, then, have the real lives of farm animals been so universally ignored over thousands of years by humans who exploit them? Why do we remain so ignorant of even the most basic knowledge about these animals? Kim Sturla gave me what is I think the correct an-

swer: because it is in our own self-interest *not* to know them; it is easier to disconnect from whom we are eating if we know nothing at all
about them. Sometimes, though, when I ask this question, I get back
astonishing answers: "Well, it's their own fault. They don't let us in
on anything. Dogs respond, cats are affectionate, but farm animals
keep their distance, they don't seem to like us very much, or to be
very interested in us, so why should I be interested in them?" Suppose, for the sake of argument, that farm animals were in fact not interested in us, that they were totally indifferent. Does this mean that
we must return the favor? Can we only be interested in an animal
who fawns on us? Is our vanity so terribly fragile that we require ado-
ration before we accord even the faintest interest? In any event, it's
not true that farmed animals are indifferent; they are frightened. It is
true that almost all farmed animals are standoffish with us, because
there is always a deep, basic, *justified* mistrust.

What is wrong, you may ask, with having a capacity for something and never exercising it? Why should it matter to us that a
chicken may be capable of deep joy but will never be able to experience it because we keep chickens in cages that make the expression of
the feeling unlikely? This is, in fact, how we define most human
tragedy: "He could have been . . ." "With her gifts . . ." What we
mean by this is that people had abilities or capacities, and something
interfered with their full development. Sometimes the fault lies in the
persons themselves; this character flaw is the subject of Greek
tragedies. When the fault lies outside the person, we have a tendency
to speak of fate intervening. We like least to talk about something
else, an outside intervention where the innate gifts, the character, the
talents, and the capabilities go unused, or unrealized, because somebody else made it impossible. Somehow this embarrasses us. We ac-

tually *prefer* ambiguity to black and white, at least theoretically, which is probably why literary critics, philosophers, and intellectuals in general seem to thrive on it. One possible reason for this is a sense of guilt at how we treat other animals whose only sin lies in being other. We don't eat each other. We cannot justify everyday cruelty to people who look just like us and seem to be put together in much the same way. But we do justify treating other living beings, farm animals, as if they were nothing but inanimate objects. They are not. It is a tragedy of our making to deprive a pig of his capacity to fully enjoy his life. We, too, become tragic, pathetic figures when we live carelessly off the suffering of other creatures.

Recently, Peter Singer has summed up the changes since the publication of his seminal *Animal Liberation*:

> These modest gains are dwarfed, however, by the huge increase in animals kept confined, some so tightly that they are unable to stretch their limbs or walk even a step or two, on America's factory farms. This is by far the greatest source of human-inflicted suffering on animals, simply because the numbers are so great. Animals used in experiments are numbered in the tens of millions annually, but last year ten *billion* birds and mammals were raised and killed for food in the United States alone. The increase over the previous year is, at around 400 million animals, more than the total number of animals killed in the US by pounds and shelters, for research, and for fur combined. The overwhelming majority of these factory-reared animals now live their lives entirely indoors, never knowing fresh air, sunshine, or grass until they are trucked away to be slaughtered.[192]

If you come away from reading this book convinced that my main thesis is correct, namely that farm animals have the capacity for all the deep feelings of their forebears, that they are remarkably similar to human beings in their ability to feel anxious, bored, sad, lonely, or deliriously happy, what are the implications?

- We should not eat meat, chicken or fish (who, recent scientific studies show, feel pain every bit acutely as do mammals).
- We should not eat eggs.
- We should not drink milk.
- We should not eat cheese, butter, cream, yoghurt, or milk chocolate.
- We should not wear leather, or wool (because shearing is unpleasant for the sheep and often hurts them), or goose down.

Becoming vegetarian was easy for me. Tofu, that wonder food, fried with onions and garlic and soy sauce, tastes better than any meat I know. Sometimes people say to me that for merely one person to become vegetarian seems futile given the magnitude of animal slaughter. "What possible good can it do for animals, if I become a vegetarian?" a friend asked me the other day. It turns out there is a good answer. Viva!—Vegetarian International Voice for Animals— states that if you become vegetarian, in your lifetime you will spare the suffering of 6 cows, 22 pigs, 30 sheep, 800 chickens, 50 turkeys, 15 ducks, 12 geese, 7 rabbits, and half a ton of fish!

I have to be honest: My research leaves me in no doubt whatsoever, that to prevent animals from suffering unbearable agony, we must become not only vegetarian, but vegan. But it has not been easy for me to make the transition from vegetarian to vegan, still a project under construction. Many years ago, when I was merely a vegetarian, I met the great Cesar Chavez, and he said to me: "If you are interested in preventing animal suffering, the first thing you should give up is eggs and milk, because the animals who produce those foods lead the most unhappy lives. You would do better to eat meat and stop eating eggs and dairy products." I was shocked, since I had no intention of eating meat but had never thought of giving up eggs or dairy products. But when I looked into it I realized he was right, and now, years later, after I have studied the matter up close, I know for certain that he was completely correct about the cruel treatment of the animals raised for such products.

The advantages of a vegan diet are enormous: for our health, for the environment, for the animals themselves. (At age sixty-two I have never been healthier than since I became semi-vegan, although exercise and plenty of outdoor activity are also responsible.) From theory to practice, however, has not been easy for me. Five years ago I gave a lecture at which I said that I was "becoming" vegan. On my last book tour, one year ago, I made the same statement, and a member of the audience jumped up and to my embarrassment reminded everybody that I had said the same thing four years previously! It is difficult. Milk has been the easiest for me to give up: soy milk actually *tastes* better than cow's milk. Even General Mills has entered the soy milk business (there are now more than fifty makers of soy milk, whereas just a few years ago there were just two). Vegetarian cheese is cheese made without rennet, a product of the stomach lining of animals, but is still made with milk. Soy cheese is a replacement, and while I find that it does not taste as good as animal cheese (my problem with cheese is not the taste, it is the ethics) it improves every year. Even Pizza Hut now serves at least one pizza with mozzarella made from soy milk, so you know the taste is becoming more acceptable to everyone. And of course why should personal taste dictate what we do, without taking into account who else must suffer? I suspect that in a few more years, soy cheese will actually taste better than goat, cow, or sheep milk cheese. Melted mozzarella from milk used to taste wonderful to me. It tastes less so, though, when I reflect on how it is made, and still less when I finally recognize the amount of suffering on the part of cows in order to produce this cheese. Why should my pleasure trump another's suffering?

Eggs have been very hard for me to give up, probably because I've loved them from such an early age. So I set about looking for true free-range eggs. Well, let me tell you, there probably is no such thing; what gets called "free range"—not a regulated term—can be just about anything from hens who truly have the run of a farm (rare) to hens who are in a barn and are never free to range anywhere outside (more common). The only truly free-range eggs are the ones to be found on a farm sanctuary, and there are still far too few such places anywhere near major cities. Many people who have thought about it even more deeply than I have, like Karen Davis, will not eat eggs even when they come from the chickens on her own sanctuary and even though they have the best life you could imagine for a chicken. She wants people to move away from the idea that their taste has a "right" to be satisfied and that animals in general, and chickens in particular, may be used to satisfy that taste. I have to agree with her.

I much admire, even envy, people who can make the intellectual connection, then take the plunge, and the next day stop eating *any animal product whatsoever*. I can't do that, though perhaps by the time this book is published, I will have at last completed my early promise. I must say, though, that anything a person does to decrease the number of animals killed is good. I used to sneer at the thought that somebody had decided to give up meat once a week. Now I think it's fine. It's a beginning. Most people who begin take the next step. Giving up red meat only is still making a move in the right direction. Giving up chicken and fish turns out to be much easier after that. (You may wake up one day and realize that the last thing you want in your stomach is a dead animal.) If the final plunge into veganism is just a month away, or several years, the desire to move in that direction has to be applauded and commended.

People who are vegetarian because they don't like to hurt animals tell me they have all kinds of tests: "I won't eat anything that runs away from me," said one. Another told me: "I don't eat what has eyes." A third woman told me she would not eat anyone who had a mother. And yesterday somebody told me he would not eat anyone with a neck and a vertebrate. I agree with all of them, and I would add: I would not eat anyone who dreams. (Did you know that bees dream? One scientist from the British Nutrition Institute who has studied the brains of bees believes "bees dream about flowers".[193]) The common ground here is that we don't want to inflict suffering, especially unnecessary suffering. We don't want to foreshorten a life. We don't want to share in the responsibility for putting animals in miserable conditions where they are forced to live their lives in totally unnatural circumstances.[194]

There probably is no such thing as a "natural" human diet. We are perhaps the only animal who has a real choice of what we can eat and remain healthy. (I should add the authority of the late Professor Stephen J. Gould, with whom I spent some time at Harvard three years ago. He told me that our dental structure suggests we are simply not carnivores. Opportunistic omnivores yes, but not an animal designed to dine primarily on meat—and I take it his view is gaining more scientific adherents all the time.) Well, how far do you go? I would say: As far as you can, without losing your sense of humor, a sense of balance, and your sanity. Raw foodists, it seems to me, go too far, at least from the point of view of animals and human health. And fruitarians, which my parents were very briefly, are verging on the charmingly nutty.

But vegetarians and vegans are not as fringe now as they were when I was growing up in California and was often the only vegetarian in my school. At least 2.5 percent of Americans are vegetarians, and about 1 percent are vegan. That may not sound like very many, but remember that means there are almost five million vegetarians in America and more than a million and a half vegans. Between the ages of twenty-five and thirty-four, the number of vegetarians is nearing 10 percent of the population. The increase in ten years is huge, and that means the food industry is paying very careful attention.

I am not going to repeat all of the very good reasons to become a vegetarian, since you can read them in several excellent books I mention in the notes to this conclusion. But I cannot refrain from pointing out that Frances Moore Lappé, in her classic *Diet for a Small Planet*, more than thirty years ago described how rearing cattle for beef was "a protein factory in reverse" since she discovered that the amount of protein, which humans could eat, fed to American livestock alone was close to the whole world's protein deficit. In other words, if Americans did not have an insatiable hunger for beef, world hunger could be eradicated. This is even truer today. Alan Durning, senior researcher at the Worldwatch Institute, believes that almost 40 percent of the world's grain and 70 percent of American grain are fed to livestock. In the developing world, Oxfam estimates that 36.1 million acres of choice land are dedicated to producing animal feeds for European livestock.

Does this mean that we have to give up the factory farm? Of course. How about the family farm? Well, yes, unless they limit themselves to growing crops. But doesn't that mean that these animals will cease to exist? We have encountered this argument earlier in the book. I need to ask: Are the people who ask this really concerned that these animals will not exist, or is it merely a ploy to stop an argument? Well, let us, for the sake of argument, suppose that somebody is worried about the extinction of farm animals. "They are only here," he would say, "because we exploit them." Would chickens really go extinct? No, they would revert to a feral existence, they would become wild again, and then their numbers would be regulated as they always have been, by ecological exigencies. Same for pigs, goats, and sheep. There would be fewer, probably, but so what? Can we really say that a chicken who is destined to a life of complete misery for a few short months is better off than never having been born? From whose point of view?

Suppose, though, that somebody says: I like farming, and I am not going to give it up. Fine, would be my response, farm vegetables, not animals. More and more people are going to want them and need them in the future. Why bring misery to animals? If, though, a farmer reads my book, but is not converted. He tells me: "I am going to raise animals. You did not change my mind. But I am interested in learning how I could treat them better. Do you have any recommendations for me?" I do not. My recommendation is to stop killing them. It just makes no sense to me that we would want to care for animals in a compassionate manner, live with them and get to know them as individuals, then turn around and simply kill them for our food. Every time I think about that, I get dizzy. It just seems profoundly wrong, as profoundly wrong as anything I know. How is it possible to care about an animal, to think about what that animal needs and wants, to think about his or her emotional life, the world of feeling he or she inhabits, and then destroy the life of that animal, another sentient being so little different than us? Author Jim Mason asks me to think about what this does to *us* as well. If we kill animals with so little concern, what is to stop us from hurting one another?

If, on the other hand, we want to live with farm animals in a kind of sanctuary situation, we would certainly learn about what they want and need. Then I would recommend environmental enrichment. The goal is to make the animals feel as much at home as possible. "Home" here refers to their ancestral environment. You want the animals to be able to realize their entire emotional potential. Again, that is an ideal we could not possibly realize, because we do not in fact know the limits of their potential (any more than we do, by the way, with a human being). They cannot, unfortunately, be turned back into their ancestors, though some could be released into "the wild." (People like to say now that they do not know what "wild" means, but it merely means not restricted by fences and other artificial devices. Some restriction is always present in nature, too, but that is a far cry from what happens in a farmyard.)

But I would take this much further than has been the case until now: the ideal way to treat a pig, today, for example, would be for the pig to live with you, as well as with other pigs. Her life must be kept interesting. You should treat her more or less the way you would treat a dog or cat. If you have a goat, never keep her alone. Goats crave the company of other goats. They need large stones to climb. As entertainers, they need to be entertained themselves, and for some reason we amuse them. If you have chickens, keep them safe, in your yard for example. Be present for them; you bring them a strange kind of security. Chickens like roosters and they want their eggs to bring forth baby chicks. They don't want to be separated from them until they are ready. And don't forget large trees where they can roost at night. If you want cows in your backyard, remember they will need other cows. And they must never be removed from their calves. Many people who live with cows have written me to say they like to hear good music, and they like to follow you as you sing to them. There are worse ways to spend your time. Sheep need a flock, and they don't like busybody dogs managing their lives. It will take time for them to trust you. You can help by supplementing their mothers' sessions by bottle-feeding them when they are young (the same is true for goats). Turkeys like to be around their wild cousins, and to be touched. The males, especially, wish to be admired for the magnificent colors on their necks and head. Ducks and geese need a real pond and the freedom to fly (it is a crime to pinion, that is, cut their flying feathers on one wing to prevent them from flying). They might leave and they might stay, but at least it will be their choice. If you must interfere with nature, keep a vigilant eye to make certain the drakes do not harm the females. All of these animals seem to enjoy being stroked. Touch is pleasurable to almost all sentient beings, even lizards!

Of course most of us are not going to live with farm animals in our backyards. So what can we do personally to help animals if the truth of this book has persuaded you, apart from becoming vegetarian and vegan?

1. Visit an animal sanctuary. There is bound to be one near you. If you live in California, go to Farm Sanctuary, not far from Sacramento, or Animal Place, just thirty-five miles from San Francisco. If you live in New York, Farm Sanctuary is near Watkins Glen, in the Finger Lakes region. If you live near the Chesapeake Bay region, or the eastern shore of Virginia, on the eastern seaboard, go to United Poultry Concerns or Poplar Springs Sanctuary, a short ride from Washington, D.C.

2. If you are a medical student, think about studying nutrition and finding ways of staying healthy without eating any animal product.

3. If you are a veterinary student, think about devoting your career to the care of farm animals.

4. If you are a graduate student of biology or zoology or psychology or any other discipline that works with animals, think about applying what you learned to farm animals.

5. If you are a librarian, make sure that the mainstream books are not the only ones you order about animals. Include some of the classic works from my list of recommended reading.

6. If you are an undergraduate at a university, organize a vegan society.

7. Join a group like the Humane Society of the United States, PETA, IDA, Compassion in World Farming, United Poultry Concerns. (See Resources.)

8. If you have money to spare, consider endowing a chair in Farm Animal Emotions, encouraging real scientists to do something really interesting for a change when it comes to animal research—non-invasive, simple observation and thought.

9. If you are already an animal welfare activist find creative ways to bring farm animals onto the agenda.

10. If you are a lawyer or law student, think of advocating legal changes that benefit farm animals.

11. If you have young children, speak to their pediatrician about making them vegan. Are you taking away their choice? Not at all; either way, you are imposing your choice on them, if you can call putting the food you eat in front of your child an "imposition."

12. If you are a farmer, refuse to work with agribusinesses that hurt animals, the environment, and you. (See below.)

13. If you work for Tyson, Monsanto, ADM, Cargill, ConAgra, Smithfield (see below), quit!

14. If I haven't yet convinced you, read more. Matthew Scully's *Dominion* is outstanding and offers a traditional case against cruelty to animals. Scully is one of the few people who has been allowed to visit an industrial hog farm. Jim Mason and Peter Singer have a new book coming out on the ethics of what you eat. In the meantime, be sure to read their *Animal Factories: What Agribusiness Is Doing to the Family Farm, the Environment and Your Health* (Harmony Books, 1990, revised edition). Still powerful, too, is Jeremy Rifkin's *Beyond Beef: The Rise and Fall of the Cattle Culture* (Dutton, 1992). Look at both books by Karen Davis, the most profound read on chickens and turkeys you will find: *More than a Meal* and *Poisoned Chickens, Poisoned Eggs*. See the film about my book, *The Emotional Lives of Farm Animals*, by the filmmaker Stanley Minasian.

15. If all this is brand new, read some of the best books in the field, the specific ones I mention in the notes, but also good general ones like Peter Singer's *Animal Liberation* (Ecco, 2001); Jim Mason's *An Unnatural Order: Why We Are Destroying the Planet and*

Each Other (Continuum, 1997); Tom Regan's *The Case for Animal Rights* (University of California Press, 1985); Mary Midgley's *Animals and Why They Matter* (University of Georgia Press, 1998). To find out about food in a more general way, read Eric Schlosser's *Fast Food Nation: The Dark Side of the All-American Meal* (Harper Collins, 2002); Marion Nestle's *Food Politics: How the Food Industry Influences Nutrition and Health* (University of California Press, 2002); John Robbins's *Diet for a New America: How Your Food Choices Affect Your Health, Happiness and the Future of Life on Earth* (1987); the wonderful novel by Ruth Ozeki, *My Year of Meats* (Viking, 1999); and if you need to know the details of a slaughterhouse, read the chilling expose by Gail Eisnitz: *Slaughterhouse* (Prometheus Books, 1997). See, too, the books in the bibliography.

16. For all of us: Stay informed. Get a good newsletter, e.g., the one put out by several organizations around farm animal issues: www.farmedanimal.net. Look at the Web sites of PETA (People for the Ethical Treatment of Animals), Compassion in World Farming, Farm Sanctuary, Animal Place, Compassion Over Killing, Mercy for Animals, and United Poultry Concerns.

17. Finally, all of us need to develop a political stance toward our food. We should not trust industry sources for our information. They have a product to sell, turf to protect, a board to satisfy with profits. They are not independent. Always consider whose interests are being served. It starts with a name: we don't hear about pigs, but pork; not cow meat, but hamburger; words designed to make us forget the origins of this food. The supermarket packaging is the same. You can rarely recognize the source.

All of this is deliberately designed to keep you from making the imaginative leap from animal suffering to the choices you make in food. Research into farm animals is generally paid for by industry. Departments of animal behavior in schools of veterinary medicine are often generously funded by the same industry. If you are doing research there, or are a graduate student, they may not forbid you outright, but you know it is not in your career interest to develop your own fully independent views if they endanger the products of your financial benefactor. How objective can we expect such research to be?

I need to say something in the end about what have been called the merchants of greed, the corporate agribusiness behemoths. Most of these names are not familiar to any of us, yet they control the world's food supply, and therefore are ultimately responsible for the suffering of billions and billions of animals every year. Cargill is the nation's largest private corporation. Archer Daniels Midland (ADM), "supermarket to the world," is the nation's single largest recipient of corporate welfare through federal subsidies and tax loopholes.[195] Tyson Foods and IBP, Inc., are the world's largest poultry producers and processor of the nation's largest meatpacking company. ConAgra, which boasts that it controls everything from "the ground to the table," is the nation's second largest food manufacturer (the first is Philip Morris—yes, the cigarette manufacturer). Smithfield Foods is the world's largest pork producer. It recently bought its putative rival, IBP, the second largest pork processor in the world. Cargill and Monsanto have joint ventures that run from fertilizer and seeds to grain and raising cattle, hogs, turkeys, and chickens, then on to the slaughterhouse. Four of these firms control 82 percent of beef, 75 percent of hogs and sheep, and half of chickens. Does this matter? Well, of course it does: every farmer who opts in is utterly dependent on a handful of agribusiness firms. In these giant companies subcontractors are often dropped, factories are closed, and workers are discarded because corporate profits (greed) dictate such inhumane gestures. If these companies treat the people who work for them in this manner, imagine what they are willing to do to the animals. Being inhumane, like a species-crossing disease, does not stop anywhere. Like independent bookstores and neighborhood groceries, the old family farm is becoming an endangered species. They are no longer farmers, but "serfs with a mortgage" as they say of themselves these days. The point is that these giant corporations are interested only in profit, for themselves and their shareholders; concerns about welfare, human or animal, just don't cut it. A.V. Krebs, who calls these eight companies "merchants of greed," points out that they will eventually destroy the family farm, do untold damage to our environment, and force farm workers into a kind of economic slavery at the same time that their behavior raises questions about the health and safety of the very food we eat.[196]

William Weida, an economist at Colorado College who counsels groups opposing hog factories, warns:

> They maintain their profitability by shifting costs off on the community. You don't put in a proper lagoon. The costs of dealing with animal waste are avoided by the owners and shifted to the surrounding population as health problems, traffic, social problems and pollution—odors, chemicals and pathogens in air or water. You don't pay the worker more than you absolutely have to. You do take advantage of every

public subsidy available. But the biggest cost issue is that hogs are a lot like humans and are sensitive to disease. That means the life of these projects is only about twelve years because the buildings become so contaminated they can't use them any longer. Too many hogs die. Then they pick up and leave, and the community is stuck with the damage.

Richard Levins, an agricultural economist at the University of Minnesota, sees farming as increasingly resembling a fast-food assembly line. "They are going to put cattle in buildings, too," he warns. According to William Heffernan of the University of Missouri, "More than 90% of all commercially produced turkey in the world comes from 3 breeding flocks. The system is ripe for the evolution of a new strain of avian flu for which these birds have no resistance. Similar concerns exist in hogs, chickens and dairy-cattle genetics."[197]

Nobody but the giant corporations themselves profit from such disregard for animals, for human health, for the environment. Farm animals have now been with us for ten thousand years. We allow them to exercise little, if any, control over their own lives. The least we owe them is to ensure that we do not destroy the very planet on which all our lives depend by shortsighted greed and indifference to human and animal suffering. We owe it to ourselves, to the animals whom we have domesticated and thereby enslaved, and to future generations to alter our views and our behavior, recognizing that we will never be able to go to sleep at night when our neighbor, human or animal, is suffering untold misery, unless we have done everything within our power to alleviate that suffering. The price for such peaceful sleep is not steep; the cost for ignoring it could be catastrophic. Let us be wise, just, and compassionate. Nothing less will do.

Afterword

A great deal has changed since I first wrote this book, especially in the UK where more than a quarter of all evening meals are now meat-free. What may have been seen as a fad, or fashion statement, or even a version of "political correctness gone mad" is now the future. Eating more fruit and vegetables, and eating less meat, poultry, and fish, is here to stay. Milk consumption is falling rapidly and you can hear the panic when milk producers protest that you cannot call plant-based milk, milk (they are fighting a losing battle, clearly, as our supermarket shelves are now stocked with soy milk, almond milk, rice milk, oat milk, and more). You can say that veganism is the "extreme" version of this new knowledge, or you can say it is the logical end point, the place where people who think about it deeply end up. If we can decrease suffering of other sentient creatures, and improve our own health, and help preserve the only planet we have, why wouldn't we? It is clearly the ethically responsible thing to do, and the UK is moving quite decisively and rapidly in that direction.

Seven percent of people in the UK are now vegan. That's three and a half million people compared to just 150,000 in 2006 (the highest number, I believe, for any country, with Israel, Italy, and Germany not far behind). Supermarket chains now sell vegan ranges alongside their mainstream offerings, vegan cookbooks soar to the top of the bestseller lists, and there are countless blogs and vlogs on the subject. And London is now Vegan Central with innumerable vegan restaurants—including London's first ever vegan pub which opened its doors earlier this year complete with animal-friendly wines and plant-based furniture.

Other countries are also moving in this direction, sometimes those you would least expect. Thus, even Italy is opening up to vegans. We were there some years ago, on a family bike tour for a week, and when I would tell the waiter "no cheese" on the pasta, he would look at me as if I came from another planet. Last year we were in Parma of all places, for a vegan festival, and there were many varieties of vegan Parmigiano and even mozzarella made from soy or nuts. We were living in Spain a short while ago, and were heartened to see that bullfighting was banned in Barcelona. In Berlin, hardly a restaurant in the hip East does not hang a sign saying "We love vegans!" Being vegetarian, they say, is so last year! Times are changing rapidly.

I am astonished at how quickly veganism seems to have advanced just about everywhere, but particularly among young people. At English universities, I have heard (though I failed to find corroboration) that at least 25 percent of undergraduates are vegetarian or vegan. Young people read, and they think, and they have strong feelings when it comes to questions of justice and equality—homelessness, racism, gender inequality, these are all immensely important to millennials, and correctly so (in these ways the world is changing for the better). They ask, quite rightly, why our concerns in these areas should be confined only to the human species. They also wonder why we confine our sympathy and our compassion to so-called domesticated animals. What is to distinguish a dog or a cat from a pig or a sheep or a cow, when it comes to their capacity for that which matters most to all of us: friendship, altruism, family ties, even love? The less interaction you have with an animal (or the more restricted that interaction is), the less complexity you ascribe to that animal. The more you know, the more intrigued (and respectful) you become. As an extreme but enchanting example, see Elisabeth Tova Bailey's popular book about a snail she befriended during an extended illness: *The Sound of a Wild Snail Eating*. There are many other examples: *H is for Hawk*, *The Soul of an Octopus*, *The Good Good Pig*, *The Secret Life of Cows*, *What a Fish Knows*.

When I wrote the first edition of *The Secret World of Farm Animals* (then called *The Pig Who Sang to the Moon*), my message was considered a very odd one. Most thought it was eccentric and many thought it a little bit crazy. I could understand, because until I began doing the research for the book, I thought being vegetarian was sufficient to reduce animal suffering. I really had to think hard to find out what was wrong with cheese and eggs and milk. They seemed so harmless. The suffering behind them was hidden from me as it was from most people. Today, that is simply not the case. More and more people recognize what goes into a glass of milk or an omelet. Is this change merely cosmetic and temporary? I seriously doubt it. The number of vegan organizations is growing by the day: in North America you have Mercy for Animals, Farm Animal Sanctuary, Physicians Committee for Responsible Medicine, the Humane Society of the United States; in Austria you have the Vegan Society Austria; in Holland the Dutch Vegan Society; in England you have Viva and the Vegan Society; in Ireland the Vegan Society of Ireland; in New Zealand you have SAFE (Save Animals from Exploitation); in Australia you have Voiceless and

Animal Liberation; to name only a few. There is no doubt that we will have "meat" grown from a dish, where no animal has to suffer, in the very near future.

I am hardly alone in believing that a major transformation has taken place and is only growing stronger. More and more people, even whole countries, now recognize the importance to our very survival of cutting back on eating food derived from animals (think, for example, of climate change and how much of it is driven by the raising of animals for food). Just as climate scientists have acknowledged this reality, medical doctors who specialize in human nutrition are also making serious changes in their recommendations for optimal health. And of course for anyone who loves their dogs and cats, it is not a huge logical leap to acknowledge that pigs and cows and sheep and chickens and ducks and rabbits all have similar sensibilities: they love their lives as much as we love ours, and they want to live without suffering too. Surely we can grant them this wish without depriving ourselves of anything essential? I have yet to meet a vegan who has not told me how much better they feel about their relationship to other animals (for we too are animals) now that they no longer kill them for food. We are, after all, the only species who gets to choose their diet. So I believe the change is permanent. There are certain things that once you know, you cannot unknow. This is one of them. And since becoming vegan is good for animals, good for you and your health and good for our planet, it is really a no-brainer, as my son would say, to get with the times.

J.M., 2018

Notes

1 Quoted in Samantha Power, *A Problem from Hell: America and the Age of Genocide* (New York: Basic Books, 2002), 42.

2 Juliet Clutton-Brock, *A Natural History of Domesticated Mammals*, 2d ed. (Cambridge: Cambridge University Press [Natural History Museum], 1999).

3 Charles Darwin, *The Expression of the Emotions in Man and Animals* (London: John Murray, 1872). See the "definitive edition" with an introduction, afterword, and commentaries by Paul Ekman, Oxford University Press, 1998.

4 George Orwell, *Animal Farm: A Fairy Story*. (Penguin edition, 1989), p. 112. It seems that nobody has noticed the importance of this quote. Thus Bernard Crick, in his definitive biography of Orwell (*George Orwell: A Life*, London: Penguin, 1982) mentions the preface, quoting, "I proceed to analyse Marx's theory from the animals' point of view" (p. 451), but omits the quotation about the exploitation of animals by humans, failing, evidently, to notice its importance. Considering that Orwell's small book is considered the greatest statement ever written about revolution, it is astonishing that Orwell's own revolutionary comment about humans and animals has been effaced from the public record!

5 Robert Wright, *NonZero: The Logic of Human Destiny* (Boston: Little, Brown, 2000).

6 The last two quotes are from I.8, 1256b 15-26; I.5, 1254b6-26.

See Richard Sorabji, *Animal Minds & Human Morals: The Origins of the Western Debate* (Ithaca, N.Y.: Cornell University Press, 1993), 135-136.

7 Quoted in Rosalind Hursthouse, *Human and Other Animals* (Oxford: The Open University, 1999), 267-9.

8 Marc D. Hauser, *Wild Minds: What Animals Really Think* (New York: Henry Holt, 2000), 253.

9 Jon Wynne-Tyson, ed., *The Extended Circle: An Anthology of Humane Thought*, 2nd rev. ed.. (Fontwell, Sussex: Open Gate Press, 1990).

10 Quoted in E. L. Grant Watson, *Animals in Splendour* (London: John Baker, 1967), 45-46.

11 Zeuner depends here on the view of the German author, Otto Antonius, who was director of the Vienna Zoo in the 1930s. His book on domestic animals is still very much worth reading: *Grundzuege einer Stammesgeschichte der Haustiere* (Jena: Gustav Fischer, 1922). See, in particular, page 241 for his full view.

12 Quoted by Richard D. Ryder: *Animal Revolution: Changing Attitudes Towards Speciesism* (Oxford: Berg, 2000), 198.

13 Karl Schwenke, *In a Pig's Eye* (Chelsea, Vt.: Chelsea Green Publishing Company, 1985), 116.

14 Quoted in F. L. Marcuse and A. U. Moore, "Tantrum behavior in the pig," *Journal of Comparative Physiological Psychology*, 37 (1944): 235.

15 His book *The Sheep-Pig*, upon which the film *Babe* was based, was first published by Victor Gollancz in 1983 and in Puffin Books in 1985.

16 Juliet Gellatley with Tony Wardle, *The Silent Ark: A Chilling Expose of Meat* —*The Global Killer* (London: HarperCollins, 1996), 9.

17 Matthew Scully, *Dominion: The Power of Man, the Suffering of Animals, and the Call to Mercy* (New York: St. Martin's Press, 2002), 260.

18 Simon de Bruxelles, "Boars on Run Keep Well-Wisher at Bay," *The Times*, January 15, 1998.

19 This episode was shown on ABC's *20/20* on September 27, 1999.

20 Alice L. Hopf, *Pigs Wild and Tame* (New York: Holiday House, 1979), 74.

21 Think about the newborn kangaroo, which is only one inch long and weighs only one-thirtieth of an ounce. But it grows up to be over six feet tall, and the mother is thirty thousand times heavier!

22 Bergljot Borresen: *Kunsten å Bli Tam* (On the Art of Being Tame) (Oslo: Gyldendal Fakta, 1994), 88. I am grateful to Dr. Borresen for telling me about this, and for translating the passage from her Norwegian book.

23 Quoted in Diana Spearman, ed., *The Animal Anthology* (London: John Baker, 1966), 69.

24 Robert Malcolmson and Stephanos Mastoris, *The English Pig: A History* (London: Hambledon Press, 1998), 130.

25 Adrian Desmond and James Moore: *Darwin* (London: Michael Joseph, 1991), 175.

26 Charles Darwin, *The Variation of Animals & Plants under Domestication*, vol. 2, 90.

27 Both quotations come from a remarkable book, the first to tell the truth about the horrible conditions that animals are raised in for food: Ruth Harrison, *Animal Machines: The New Factory Farming Industry*, with a foreword by Rachel Carson (London: Vincent Stuart, 1964), 97.

28 Quoted in Karl Jacoby, "Slaves by Nature? Domestic Animals and Human Slaves," *Slavery and Abolition* 15 (1994): 89-99.

29 Ruth Harrison, "Ethical Questions on Modern Livestock Farming." In David Paterson and Richard D. Ryder, eds., *Animals' Rights: A Symposium* (London: Centaur Press, 1979), 125.

30 F. E. Zeuner, *A History of Domesticated Animals* (London: Hutchinson, 1963), 266.

31 Wesley Mills, *The Nature and Development of Animal Intelligence* (London: T. Fisher Unwin, 1898), 40.

32 Temple Grandin, *Meat and Poultry*, May 1992. Quoted in *Battered Birds, Crated Herds: How We Treat the Animals We Eat*. Gene Bauston: Farm Sanctuary, 1996.

33 Jeffrey Moussaieff Masson, *The Emperor's Embrace: Fatherhood in Evolution* (New York: Pocket Books, 2000).

34 Thomas Hardy, in *Jude the Obscure*, noted this same look of deep reproach in a pig that was horribly slain: "The dying animal's cry assumed its third and final tone, the shriek of agony; his glazing eyes riveting themselves on Arabella with the eloquently keen reproach of a creature recognizing at last the treachery of those who had seemed his only friends." Quoted in F. C. Sillar and R. M. Meyler, *The Symbolic Pig: An Anthology of Pigs in Literature and Art* (Edinburgh and London: Oliver & Boyd, 1961), 160.

35 Charles Darwin, *The Variation of Animals and Plants under Domestication* (London: John Murray, 1905), chap. 13, p. 8. In chapter 3, on domestic pigs (p. 93) he writes: "It is a remarkable fact that the boars of all domesticated breeds have much shorter tusks than wild boars. Many facts show that with many animals the state of the hair is much affected by exposure to, or protection from, climate . . . may we not venture to surmise that the reduction of the tusks in the domestic boar is related to his coat of bristles being diminished from living under shelter? On the other hand, as we shall immediately see, the tusks and bristles reappear with feral boars, which are no longer protected from the weather."

36 Hursthouse, *Humans and Other Animals*, 267-9.

37 Joyce D'Silva, the head of Compassion in World Farming, an advocacy group in England that has proven remarkably successful in changing laws concerning farm animals in Europe, gave me this information.

38 From a BBC video on clever pigs called *Move over Babe*, a Creative Touch and Laurel Production, 1997.

39 See the entry "Pigs," written by Andrew Vayda, in Peter Ryan (ed.), *Encyclopaedia of Papua and New Guinea* (Melbourne: Melbourne University Press, 1972), vol. 2, 905-9. On page 906, Vayda writes: "Because of this pervasive interest in pigs, some New Guinea people have been characterized as having a 'pig complex' comparable to the 'cattle complex' of East African pastoralists. The preoccupation is expressed succinctly in the Western Highlands District of Australian New Guinea by the Enga men who say, 'Pigs are our hearts!' The place of pigs in the social and religious life of another group of Highlanders, those in

the vicinity of Nondugl in the Middle Wahgi, is said to be such that it cannot be filled by any amount of money, wives, or bird of paradise plumes.

40 Among these people, as among others, the women, to whom the care of pigs is usually entrusted, take pride in the animals, give names to them, sleep in the same house with them, and fondle them when small. Sometimes, in the Highlands, a woman even allows orphaned piglets or those of large litters to share her breasts temporarily with her children. Solicitous care of pigs may take other forms elsewhere, but it is widespread in one form or another in lowland as well as highland regions. For example, among the Siuai people of the Greater Buin Plain of Bougainville, young pigs are cared for as if they were pets; their food is cooked in one pot with their owners' food; the women premasticate tubers for sickly piglets; and the animals are ritually named, 'baptized,' and given magical treatment for ailments. Even in the difficult Tor region or West Irian, where the people's wandering from one sago grove to another is not conducive to the breeding of pigs, there is something like a pig complex. Here wild piglets are sometimes caught and raised in the villages, where they are fed with sago, provided with names, and addressed by kinship terms. When grown, the animals are killed and eaten but not by the households that have been rearing them, for, as Tor people say, 'Who would eat his own son or brother?'"

41 Valerie Porter, *Pigs: A Handbook to the Breeds of the World* (Sussex: Helm Information, 1993), 199-200.

42 Heinz Meynhardt, *Schwarzwild-Report: Mein Leben unter Wildschweinen* (Leipzig: Neumann Verlag, 1986).

43 Tamara Staples and Ira Glass, *The Fairest Fowl: Portraits of Championship Chickens* (San Francisco: Chronicle Books, 2001).

44 Charles Darwin, *The Descent of Man and Selection in Relation to Sex*, 2nd ed. (London: John Murray, 1874), 129. The work referred to is *Facultes mentales des animaux comparée à celles des hommes* (Paris: 1872), in two volumes. See note 137 where I discuss this work, to which Darwin often refers.

45 An expert on the sounds made by chickens, the animal behaviorist Domhnall Jennings, writes me: "Chickens are a non-songbird

species. What makes them such a challenge to study is the large range of calls that they have and that these seem to be pre-programmed (other non-songbirds such as the dove have 3 different calls). It is still not clear how many calls chickens have since we can't be sure whether calls are grading into each other and as such are not entirely distinct from each other. I would tend to go along with Collias and plump for about 24 and this includes juvenile and adult (male and female) calls."

46 William Grimes, *My Fine Feathered Friend* (New York: North Point Press, 2002), 64.

47 See the chapter "Domestic Fowl" in Zeuner, *A History of Domesticated Animals.*

48 J. D. Elius, "Sapient Sauropsids and Hollering Hominids." In W. Koch (ed.), *Geneses of Language* (Brockmeyer: Bochum, 1987), 11-29.

49 Lesley J. Rogers, *The Development of Brain and Behaviour in the Chicken* (Oxford: CAB International, 1995), 221.

50 Page Smith and Charles Daniel, *The Chicken Book* (San Francisco: North Point Press, 1982), 177.

51 Quoted in Hauser, *Wild Minds.*

52 Rogers, *The Development of Brain and Behaviour in the Chicken,* 215.

53 Karen Davis, *Prisoned Chickens Poisoned Eggs: An Inside Look at the Modern Poultry Industry* (Summertown, Tenn.: Book Publishing Company, 1996), 34. This book gives a great deal of information and is written by one of the best-informed people about chickens in the United States.

54 Zdenek Veselovsky, *Are Animals Different?* Translated from the German by Anne Rasa, edited by Michel Boorer (London: Methuen, 1973), 207-8.

55 Irene Maxine Pepperberg, *The Alex Studies: Cognitive and Communicative Abilities of Grey Parrots* (Cambridge: Harvard University Press, 1999), 327. Note that Darwin, in *The Descent of Man*, p. 130, wrote that "it is certain that some parrots, which have been taught to speak, connect unerringly words with things, and persons with events."

56 Razvan A. Tuculescu and Joseph Griswold, "Prehatching Interaction in Domestic Chickens," *Animal Behaviour* 31 (1983): 1-10. Of

course it was known for some time that sounds were communicated back and forth from parent to embryo. Veselovsky wrote that "when I hatched a rhea's egg in an incubator I found that it was possible to detect movement inside the egg several days in advance. Later, when I held the egg and said a few words, a long, whining call came from it. The male rhea would answer this with a call which has a calming effect." Op cit., p. 117. He points out that this kind of communication only happens in nidifugous birds, those who can run as soon as they are hatched. The reason is that they need to be able to recognize their parents when they immediately separate from them at birth.

57 Alice Walker, *Living by the Word: Selected Writings 1973-1987* (San Diego: Harcourt Brace Jovanovich, 1988), 171.

58 Darwin, *The Descent of Man*, p. 106. Darwin gives examples, one from Rengger who "observed an American monkey (a Cebus) carefully driving away the flies which plagued her infant; and Duvaucel saw a Hylobates [gibbon] washing the faces of her young ones in a stream." Note that this is from the second edition, which contains many important additions. It would be interesting to know why Darwin added so much (nearly double in the important chapter 3, "Comparison of the Mental Powers of Man and the Lower Animals"), but I know of no study addressing this issue. William Whewell (1794-1866) was the first exponent of the philosophy and history of science. Johann Rudolph Rengger (1795-1832) was a Swiss medical doctor who was in Paraguay in 1819 and wrote a book about the mammals of that country, which Darwin often quoted from with approval. Darwin taught himself German to be able to read the scientific literature in that language, but often complained how difficult it was for him to read. The amount of quotations from German literature is nonetheless astounding.

59 Romanes, op. cit., 317.

60 Christine J. Nicol and Stuart J. Pope: "The Maternal Feeding Display of Domestic Hens is Sensitive to Perceived Chick Error," *Animal Behaviour* 52 (1996): 767-74. See, by the same authors, "Social Learning in Small Flocks of Laying Hens," *Animal Behaviour* 47: 1289-96.

61 T. C. Danbury et al., "Self-Selection of the Analgesic Drug Carprofen by Lame Broiler Chickens," *The Veterinary Record*, March 11, 2000, 307-11.

62 Aristotle, *History of Animals*, book VIII (631b-xlix), edited and translated by D. Balme, prepared for publication by Allan Gotthelf (Cambridge: Harvard University Press, 1991), 399. This passage was quoted by George John Romanes, *Mental Evolution in Animals*, in the chapter "Imperfection of Instinct," p. 171, where he quotes Pliny: "He did everything for them, like to the very hen that hatched them and ceased to crow." He goes on to quote other authors to the same effect. But I have not heard of this happening in modern times. Are we missing something or did the older writers imagine it?

63 *Aldrovandi on Chickens: The Ornithology of Ulisse Aldrovandi (1600), Vol. II, Book XIV*, translated from the Latin with introduction, contents, and notes by L. R. Lind (Norman, University of Oklahoma Press, 1963).

64 Valerie Porter: *Domestic and Ornamental Fowl* (London: Penguin [Pelham Books], 1989), 84.

65 There is a tonic immobility test, in which a bird goes into a catatonic-like state when restrained on its back. See G. G. Gallup, "Tonic Immobility as a Measure of Fear in Domestic Fowl," *Animal Behaviour* 27 (1979): 316-317. But we should remember that some chickens are so trusting, and get such pleasure from being stroked, that they may, much like a kitten, roll over playfully to elicit caresses. One's chicken terror may be another's pleasure.

66 Juliet Clutton-Brock: *A Natural History of Domesticated Mammals*, 2d ed. (Cambridge: Cambridge University Press [Natural History Museum]), 1999, p. 14.

67 I am not sure where this silly urban legend originated, but it continues to be repeated with tiring frequency, even by people who should know better. Thus William Grimes, in an article in the *New York Times* "Week in Review" (January 12, 2003) speaks of turkeys as being the gold standard for stupidity. "While chickens can survive a rainstorm outside, turkeys will look skyward and drown as their throats fill with water." Karen Davis, who know more about turkeys than anyone,

thinks the myth may come from the fact that baby chicks, separated from their mothers and without the protection of her wings in a storm, may well look up, inquiringly or beseechingly, accidentally fill their small beaks with water, and drown. Like so much of what we consider to be stupidity, however, this is a human artifact, just one more example of our cupidity. No turkey chick should ever have to grow up without a mother. In the wild, turkeys live with their mothers for many months before striking out on their own. By then they have learned to live in a dangerous world.

68 Joe Hutto, *Illumination in the Flatwoods: A Season with the Wild Turkeys* (New York: Lyons & Burford, 1995).

69 In November 2002, Oklahoma voters banned cockfighting in Oklahoma in a ballot initiative.

70 J. G. Romanes, *Mental Evolution in Animals: With a Posthumous Essay on Instinct by Charles Darwin* (London: Kegan Paul, Trench & Co., 1883), 381-2. Romanes (1848-1894) was a biologist who worked at University College, London, and later Oxford. He was a close friend of Darwin from 1874 onwards.

71 F. Galton, "The First Steps Towards the Domestication of Animals," *Transactions of the Ethological Society of London*, NS 3 (1865): 122-38; the quote is from p. 124.

72 J. P. Kruijt, *Ontogeny of Social Behaviour in Burmese Red Junglefowl (Gallus gallus spadiceus) Bonnaterre* (Leiden: E. J. Brill, 1964), 9.

73 Marian Stamp Dawkins, *Through Our Eyes Only? The Search for Animal Consciousness* (Oxford: Oxford University Press, 1998), 153.

74 N. E. Collias and E. C. Collias, "A Field Study of the Red Jungle Fowl in North-central India," *Condor* 69 (1967): 377.

75 I already referred to her book *Prisoned Chickens, Poisoned Eggs*. See too her "Viva, The Chicken Hen," *Between the Species: A Journal of Ethics*, 6.1 (1990), 33-35. Also her article "Thinking Like a Chicken: Farm Animals and the Feminine" in Carol J. Adams and Josephine Donovan, eds., *Animals and Women: Feminist Theoretical Explorations* (Durham: Duke University Press, 1995). See the Web site of United Poultry Concerns (PO Box 150, Machipongo, VA 23405): www.upc-online.org.

76 Collias and Collias, "A Field Study of the Red Jungle Fowl in North-central India," 360-86.

77 Gilbert White, *The Natural History of Selborne*, edited with an introduction by R. M. Lockley (New York: Dutton [Everyman's Library], 1949), 187.

78 Porter, *Domestic and Ornamental Fowl*, 184.

79 Marian Stamp Dawkins, *Animal Suffering: The Science of Animal Welfare* (London: Chapman & Hall, 1980), 114.

80 From the introduction to Clare Druce, *Chicken & Egg: Who Pays the Price?* (London: Merlin Press, 1989).

81 David Tomlinson, *Ducks* (London: Whittet Books, 1996), 49.

82 D. G. M. Wood-Gush, *The Behaviour of the Domestic Fowl* (London: Heinemann, 1971). (Rpt. 1989.)

83 D. G. M. Wood-Gush, Ian J. H. Duncan, and C. J. Savory, "Observations on the Social Behaviour of Domestic Fowl in the Wild," *Biology of Behaviour* 3 (1978), 193-205.

84 See www.grandin.com/welfare/corporation.agents.html. The paper is called "Corporations Can Be Agents of Great Improvements in Animal Welfare and Food Safety and the Need for Minimum Decent Standards."

85 Professor Lesley Rogers acknowledges that the present living conditions for chickens cannot possibly "meet the demands of a complex nervous system designed to form a multitude of memories and to make complex decisions," op cit. p. 218. Key to my thesis about chickens is the truth of this statement. What memories can a chicken who has been confined to a battery cage possibly have? This understimulation of the brain is one more crime we have committed against the chicken.

86 Douglas H. Chadwick, *A Beast the Color of Winter: The Mountain Goat Observed* (San Francisco: Sierra Club Books, 1983), 80-81. For more on wild sheep, see George B. Schaller's *Mountain Monarchs: Wild Sheep and Goats of the Himalaya* (Chicago: University of Chicago Press, 1975), and Valerie B. Geist, *Mountain Sheep: A Study in Behavior and Evolution* (Chicago: University of Chicago Press, 1971).

87 Elizabeth Arthursson: *Ewes & I* (London: Souvenir Press, 1988), 106.

88 Nathaniel Southgate Shaler, *Domesticated Animals: Their Relation to Man and to his Advancement in Civilization* (New York: Charles Scribner's Sons, 1895).

89 Euan Macphail, *The Evolution of Consciousness* (Oxford: Oxford University Press, 1998) and Stephen Budiansky, *If a Lion Could Talk: Animal Minds and the Evolution of Consciousness* (New York: Free Press, 1999).

90 See K. M. Kendrick et al., "Facial and Vocal Discrimination in Sheep," *Animal Behaviour*, 49 (1995): 1665-76.

91 P. D. Morgan and G. W. Arnold, "Behavioural Relationships between Merino Ewes and Lambs During the Four Weeks After Birth," *Anim. Prod.* 19 (1974), 196.

92 Possibly the oldest of all domestic sheep breeds is the *karakul* in Bukhara, Uzbekistan (from where my father's family comes). There are about 4 million sheep being farmed there for their pelts. The baby lambs, as soon as they are born, are slaughtered for their lustrous, glossy, curly black skin, which is like crushed velvet. This is known as karakul fur and is extremely sought after by fashion designers. Even more so is what is known as broadtail (another name for the sheep) which refers to the skin of the unborn lamb, two weeks before birth, which is even softer. This is also called Persian lamb or Astrakhan, and is used by most of the big designers, including Versace, Fendi, Christian Dior, Prada, Karl Lagerfeld, and Ralph Lauren, and is sold by Macy's, Bloomingdale's, Neiman Marcus, and Saks Fifth Avenue. It takes at least thirty pelts to make a coat. A broadtail outfit sells for upward of $25,000. The Humane Society of the United States went to Bukhara and in March 2001 did a thorough investigation. It is appalling. The killing is done with absolutely no concern for the animals. The report can be found on their Web site, www.hsus.org. On the PETA Web site (www.peta.org) is an expose, entitled "Inside the Wool Industry," which points out that many designers are now using Shahtoosh shawls (they sell for $15,000 each) which come from the Endangered Tibetan antelope, the *chiru*.

This antelope cannot be domesticated, and must be killed to obtain the wool. There are cheap and wonderful wool substitutes, Tencel and Polartec Wind Pro, for example, and there is no excuse to cause this amount of animal suffering for human vanity.

93 Jaak Panksepp: "Cats: The Chemistry of Caring" in *Smile of a Dolphin: Remarkable Accounts of Animal Emotions*, with a foreword by Stephen Jay Gould (New York: Discovery Books, 2000), 58.

94 See Bruno Auboiron and Gilles Lansard, *La transhumance et le berger: une tradition vivante* (Aix-en Provence: Édisud, 1998), 100.

95 See also Nicole Reynes and Christophe Latour, *Moutons & bergers* (Paris: Editions Rustica, 2000).

96 No doubt milking ewes goes far back in history, but the first documentation of it happening is not until the ninth century. As wool became increasingly valuable, milking sheep declined as it was recognized that lambs deprived of their mother's milk do not have the same quality of fleece. See Judy Urquhart, *Animals on the Farm: Their History from the Earliest Times to the Present Day* (London: Macdonald & Co., 1983), 51, for many valuable facts about sheep.

97 I did not know this, but I take the information from Urquhart, *Animals on the Farm*, 62.

98 Zeuner, *A History of Domesticated Animals*, 170.

99 See his account of the training of sheep dogs in *The Domestic Dog: An Introduction to Its History* (London: Routledge and Kegan Paul, 1957), 128-136.

100 Janet White, *The Sheep Stell: The Autobiography of a Shepherd* (Herts: The Sumach Press, 1991), 190.

101 See Yutaka Tani, "The Geographical Distribution and Function of Sheep Flock Leaders: A Cultural Aspect of the Man-Domesticated Animal Relationship in Southwestern Eurasia." In Juliet Clutton-Brock, ed., *The Walking Larder: Patterns of Domestication, Pastoralism, and Predation* (London: Unwin Hyman, 1989), 181.

102 *Aelian on the Characteristics of Animals*, with an English translation by A. F. Scholfield, 3 vols. (London: William Heinemann, 1958). The passage is V, 48 (p. 345).

103 Quoted (from an 1835 book on fossil fuel by J. Holland), in Zeuner, *A History of Domesticated Animals*, 198.

104 Germaine Greer, "Who Says Sheep Are Senseless? (Country Notebook)," *Daily Telegraph*, June 12, 1999, Country, 11.

105 Charles G. Hansen, "Sense and Intelligence." In Gale Monson and Lowell Sumner, eds., *The Desert Bighorn: Its Life History, Ecology & Management* (Tucson: University of Arizona Press, 1980).

106 Norman M. Simmons, "Behavior." In Monson and Sumner, *The Desert Bighorn*, 134.

107 Colin Tudge, "Farms in Loco Parentis," *New Scientist*, Oct. 18, 1973, 179-181.

108 E. S. E. Hafez., ed., *The Behaviour of Domestic Animals*, 3d ed. (London: Bailliere Tindall, 1975), 38.

109 Clearly this is the argument that Colin Tudge makes, for he believes that the matter is an aesthetic one, rather than an ethical one. In the article quoted in the preceding note he goes on to say, "If it is finally concluded that the appropriate breed of chicken in the appropriate cage is a reasonably happy creature, then the merely aesthetic point that we do not like to see animals behind bars, becomes irrelevant." The point, surely, is that we must seek to answer such questions of animal happiness from the point of view of the animal, not our own. Until we can demonstrate conclusively that wild sheep would prefer to live in human confines, we cannot claim they are happier with us than on their own. The only relevant argument is not whether we do not like to see animals behind bars, but the preferences of the animals themselves. For humans the point is an ethical one; only for the animals could it possibly be aesthetic, though I suspect it is far more urgent than that.

110 London; Weidenfeld & Nicholson, 1994, p. 167.

111 See the useful collection *Evolution of Domestic Animals*, edited by Ian L. Mason (London: Longman, 1984).

112 Charles Darwin, *The Variation of Plants and Animals under Domestication* (1868, reprint, foreword by Harriet Ritvo, Baltimore: Johns Hopkins University Press, 1998), vol. 2, p. 331. "According to two accounts, the *Hypericum crispum* in Sicily is poisonous to white sheep

alone; their heads swell, their wool falls off, and they often die." Darwin says that these cases "are at present wholly inexplicable."

113 A. J. F. Webster, "Animals and Husbandry," Annual Lecture of the Royal Agriculture Society of England, 1997.

114 We should not forget that these diseases are primarily man-made. Foot rot, for example, is inseparably linked with the environment, caused by the animal standing in wet conditions for too long. Virulent foot rot is extremely painful to the sheep; affected animals stand on their knees or limp severely. The disease may continue for a year or more. See the outstanding book by Christine Townend, *Pulling the Wool: A New Look a the Australian Wool Industry* (Sydney: Hale & Ironmonger, 1985).

115 Peter Isaac, ed., *The Farmyard Companion* (London: Jill Norman & Hobhouse, 1981), 130.

116 C. V. Hulet, G. Alexander, and E. S. E. Hafez, "The Behaviour of Sheep." In E. S. E. Hafez, ed., *The Behaviour of Domestic Animals*, 3rd ed. (London: Bailliere Tindall, 1975), 252.

117 Urquhart, *Animals on the Farm*, 71. The book by William Youatt was published by Simpkin, Marshall & Co. in 1878, and was entitled *Sheep, Their Breeds, Management, & Diseases.*

118 J. P. Scott, "Social Behavior, Organization and Leadership in a Small Flock of Domestic Sheep." In *Comparative Psychology Monographs*, vol. 18 (Maryland: William & Wilkins, Co., 1945), 26.

119 The word *kid* comes from the Old Norse *kio*, the young of a goat. In the seventeenth century it came to mean a human child, but only as slang. Today it is standard English. In early nineteenth-century England *to kid* is to make a goat of someone, hence our use of the word to mean to tease or make fun of.

120 Zeuner, *A History of Domesticated Animals*, 135.

121 *The Wordsworth Dictionary of Phrase & Fable* (London: Wordsworth Editions, 1993), s.v.

122 C. J. Stevens, *One Day with a Goat Herd* (Phillips, Me.: John Wade, 1992).

123 Urquhart, *Animals on the Farm*, 20.

124 Peter Lovenheim, *Portrait of a Burger as a Young Calf: The Story of One Man, Two Cows, and the Feeding of a Nation* (New York, Harmony Books, 2002).

125 Laurie Winn Carlson, *Cattle: An Informal Social History* (Chicago: Ivan R. Dee, 2001).

126 Quoted in Oliver W. Sacks's *An Anthropologist on Mars*, 1996.

127 Quoted in Rebecca Hall, *Voiceless Victims* (London: Wildwood House, 1984).

128 This is from Moses Maimonides, *Guide for the Perplexed*, pt. 3, chap. 48. Translated by M. Friedlander, Dover, 1950.

129 This is quoted in a book by Al-Hafiz B. A. Masri, who was the Imam of the Shah Jehan mosque in Woking, in a book called *Animals in Islam* (Petersfield: Athene Trust, 1989).

130 Quoted in Carlson: *Cattle: An Informal Social History*, 31.

131 W. Youatt, *Cattle, Their Breeds, Management, & Diseases* (London: Baldwin and Cradock), 1838.

132 M. Kiley-Worthington, *The Behavioural Problems of Farm Animals* (Stockton: Oriel Press, 1977).

133 Robert Schloeth, "Das sozialleben des Camargue-Rindes," *Zeitschrift fur Tierpsychologie,* 18 (1961): 530-627.

134 The first quote is from Ramayana 2.68.15-25, and the second is from 2.17.32–33 of the critical edition of the epic text.

135 M. K. Gandhi, *How to Serve the Cow* (Ahmedabad: Navjivan, 1954).

136 Darwin: *The Descent of Man and Selection in Relation to Sex*, 2nd ed., 156.

137 There is no footnote in Darwin's text, but I am certain that the passage Darwin had in mind is from the book *Études sur les facultés mentales des animaux comparées à celles de l'homme* (Studies on the Mental Abilities of Animals Compared to Humans) of 1872. The first volume is anonymous, saying only that it was written by "un voyager naturaliste," but in the second volume his name is given as J. C. Houzeau. The passage is found on page 471 of the second volume.

138 It did not occur to Houzeau or to Darwin that the compassion

whose absence they lament in the cow is also not forthcoming on the part of humans toward cows. We slaughter them in astronomical numbers every year. An exception to the general indifference and lack of sympathy with cows is the altogether remarkable, long short story by the American author James Agee called "A Mother's Tale," which was published in *Harper's Bazaar* in 1952 and reprinted in Robert Fitzgerald, ed., *The Collected Short Prose of James Agee* (Boston: Houghton Mifflin, 1968), 221-43. It is told from the point of view of a young calf who suddenly realizes that the trains he sees are carrying cattle to slaughter. "I'm told that far back in the wildest corners of the range there are some of us [cows], mostly very, very old ones, who have never been taken. It's said that they meet, every so often, to talk and just to think together about the heroism and the terror of two sublime Beings. The One Who Came Back, and The Man With The Hammer."

139 D. A. Shutt, et al, "Stress Induced Changes in Plasma Concentrations of Immunoreactive B Endorphin and Cortisol in Response to Routine Surgical Procedures in Lambs." *Australian Journal of Biological Science* 40 (1987): 97-103.

140 John Webster, *Animal Welfare: A Cool Eye Towards Eden* (Oxford: Blackwell, 1995), 93.

141 See his own account: Lutz Heck, *Animals, My Adventures* (London: Methuen, 1954).

142 Edward Hyams, *Working for Man: The Domestication of Animals* (Harmondsworth, Middlesex: Penguin Books, 1975).

143 See S. S. Campbell and I. Tobler, "Animal Sleep: A Review of Sleep Duration across Phylogeny," *Neuroscience and Bio-behavioral Reviews* 8 (1984): 269-300. Also, H. Zepelin and A. Rechtschaffen, "Mammalian Sleep, Longevity, and Energy Metabolism," *Brain Behavior and Evolution* 10 (1974): 425-70.

144 *L'Equitaine, le Cheval et l'Ethologie* (Paris: Belin, 1999).

145 Darwin, in *The Descent of Man*, writes, "It deserves notice that the instinct of pairing with a single female is easily lost under domestication. The wild-duck is strictly monogamous, the domestic-duck

highly polygamous." I have used the edition (originally published in 1868) edited by Harriet Ritvo, Johns Hopkins University Press, 1998, part 2, p. 270. Darwin mentions this several times in volume 2 of *The Variation of Animals and Plants under Domestication,* first published in 1868 and revised in 1883, so it was clearly an important point for him. It is interesting that all animals we domesticate tend to lose their natural sexuality, whereas some wild animals that we keep in zoos do the opposite: they refuse to propagate at all.

146 Ted Hughes, *What is the truth? A farmyard Fable for the Young,* vol 2 (London: Faber, 1984).

147 This and other charming passages are to be found in the lovely book by Arthur Cleveland Bent, *Life Histories of North American Wild Fowl,* originally published in two volumes in 1923 and 1925 by the Smithsonian Institution, and reprinted by Dover in 1987. See, too, the examples and the explanation given by Edward A. Armstrong under the heading "distraction display" in his fine book *Bird Display and Behaviour: An Introduction to the Study of Bird Psychology* (1942, reprint, New York: Dover, 1965).

148 Konrad Lorenz and Nikolaas Tinbergen, "Taxis and Instinctive Action in the Egg-Retrieving Behaviour of the Greylag Goose." In C. H. Schiller and K. S. Lashley, eds., *Instinctive Behaviour* (New York: International University Press, 1957).

149 Bryce Fraser, *Sitting Duck: A True Story* (London: MacGibbon & Kee, 1971).

150 Theodore Xenophon Barber, *The Human Nature of Birds: A Scientific Discovery with Startling Implications* (New York: St. Martin's Press, 1993), 115.

151 Quoted in John Robbins, *Diet for a New America* (Walpole, N.H.: Stillpoint Publishing, 1987), 24.

152 Randy Thornhill and Craig Palmer, in *A Natural History of Rape: The Biological Basis of Sexual Coercion* (Cambridge: M.I.T. Press, 2001), p. 145 mention mallards, but they misinterpret the leading authority on forced copulation in ducks, the evolutionary biologist and

feminist Patricia Gowaty. See her article, with N. Buschhaus, "Ultimate Causation of Aggression and Forced Copulation in Birds," *American Zoologist* 38 (1988): 207-25.

153 F. McKinney, "The Behaviour of Ducks." In Hafez, *The Behaviour of Domestic Animals*, 252.

154 The scientific arguments can be found well explained in the brilliant book by Joel Carl Welty, *The Life of Birds* (New York: Knopf, 1968).

155 William Lishman, *Father Goose: The Adventures of a Wildlife Hero* (Boston: Little, Brown, 1995), 151.

156 See the section on domestication in Janet Kear, *Man and Wildfowl* (London: T & A. D. Poyser, 1990), 22-70.

157 See M. J. Gentle and L. N. Hunter, "Physiological and Behavioural Responses Associated with Feather Removal in *Gallus var domesticus*," *Research in Veterinary Science* 50 (1990): 95-101.

158 A useful source of all information on farm animals (completely lacking, however, in any perspective on the rights of these animals), particularly ducks and geese, is the collection edited by Gail Damerow, *Barnyard in Your Backyard: A Beginner's Guide to Raising Chickens, Ducks, Geese, Rabbits, Goats, Sheep, and Cattle* (North Adams, Mass.: Storey Books, 2002).

159 I have taken my information about down from an article by Elliot L. Gang, "Down with Down," *Animal's Agenda Magazine*, May/June 1998. Here is what he writes: "Workers in Eastern European countries, the source of most commercially available down, rip the feathers by hand from the necks and breasts of live, fully conscious geese and ducks after tying their feet together and hanging them upside down. Afterwards, traumatized and in agony, the birds tremble uncontrollably and huddle together or find a structure to lean on for support in their shock. It takes them several days to recover. Workers allow the birds' feathers to grow back and subject them to the same ordeal four or five times in their lives." I am sure this is an eyewitness account, but I cannot vouch for how common this is, or whether it goes on everywhere.

160 On live plucking, I took this information from PETA *News*,

Nov.-Dec. 1988. For the factory farming of ducks, see Compassion in World Farming's 2001 report "The Factory Farming of Ducks," which is on their Web site (www.ciwf.co.uk). Confined ducks are also subject to other painful eye diseases from excretory ammonia fumes from the decomposing uric acid in their droppings. See the excellent letter to the EPA by United Poultry Concerns' Karen Davis in January 2002, which can be found on its Web site (www.upc-online.org). I should also mention that the cages ducks are confined to have wire floors, which is very hard on their webbed feet.

161 There is an interesting account, in the classic *Bewick's British Birds*, vol. 2, of how down was collected in the eighteenth century: "In Greenland, Iceland, Spitzbergen, Lapland, and some parts of the coasts of Norway, the Eiders flock together, in particular breeding places, in such numbers, and their nests are so close together, that a person in walking along can hardly avoid treading upon them. The natives of these cold climates eagerly watch the time when the first hatchings of the eggs are laid: of these they rob the nest, and also of the more important article, the down with which it is lined, which they carefully gather and carry off. These birds will afterwards strip themselves of their remaining down, and lay a second hatching, of which also they are sometimes robbed: but it is said, that when this cruel treatment is too often repeated, they leave the place and return to it no more." This is the famous Thomas Bewick (1753-1824) and the book is his *Water Birds*, published in 1804.

162 H. Albert Hochbaum, *Travels and Traditions of Waterfowl* (Minneapolis: University of Minnesota Press, 1955), 54.

163 Derek Tangy, *A Drake at the Door* (London: Michael Joseph, 1963), 139.

164 Darwin, *The Descent of Man*, part 2, p. 110.

165 This is according to the world authority Jean Delacour, *The Waterfowl of the World*, vol. 1. (London: Country Life, 1954), 94. By the time Delacour was 15, in 1908, he had the largest collection of living wildfowl in the world (it was destroyed in World War I, rebuilt, then destroyed again during World War II).

166 W. H. Hudson, *Birds and Man* (London: Duckworth, 1924), 211.

167 People for the Ethical Treatment of Animals (PETA) in 1991 investigated foie gras production at Commonwealth Enterprises, located in the Catskills of New York. Here is their report, from their Web site: "Despite Commonwealth's many prior claims that it made foie gras without force-feeding the ducks, PETA's investigators observed and documented the following: Three times a day, workers entered small duck pens in a factory-farm building. The ducks, knowing what was coming, struggled to get as far away from the men as possible. The workers grabbed the ducks one at a time, held them down, forced open their bills, and shoved a long metal pipe down their throats all the way to their stomachs. They then squeezed a lever attached to the pipe, and an air-driven pump forced a third of the day's six-to-seven pounds of corn mixture into each duck's stomach. Each worker was expected to force-feed 500 birds three times a day. So many ducks died when their stomachs burst from overfeeding that workers who killed fewer than 50 of 'their' 500 received bonuses. After four weeks of force-feeding, the ducks were slaughtered, their livers six to twelve times normal size—pale, blotchy melon-sized messes instead of small, firm, healthy organs."

168 Keith Thomas, *Man and the Natural World: A History of the Modern Sensibility* (New York: Pantheon, 1983), 94.

169 Geese, which are usually kept in small pens or cages, now account for less than 5 percent of foie gras production in France. I am not certain why this is the case, perhaps because geese are larger and more difficult to handle? In 1998, 24,270,000 ducks were force-fed in France, compared to just 638,000 geese. These figures come from the inquiry mentioned in the next note.

170 Carol McKenna, *Forced Feeding: An Inquiry into the Welfare of Ducks and Geese Kept for the Production of Foie Gras* (London, World Society for the Protection of Animals, 2000). An excellent article is "Ducks out of Water," a report on the duck industry in the United States by

Juliet Gellatley (with research by Lauren Ornelas). This can be found online at www.vivausa.org.

171 McKenna, *Forced Feeding*, 15. The first quotation is from Wendy Jensen, D.V.M.

172 For example, the *Larousse Gastronomique*, or M. Toussaint-Samat's *History of Food* (Blackwell, 1992).

173 Michael Meyer, *Henrik Ibsen* (New York: Viking, 1987).

174 Darwin, *The Variation of Animals and Plants under Domestication*, vol. 1, p. 293.

175 Roger Scruton, *Animal Rights and Wrongs* (London: Demos, 1996), 141.

176 Ian J. H. Duncan and David Fraser, "Understanding Animal Welfare." In Michael C. Appleby and Barry O. Hughes, eds., *Animal Welfare* (Oxford: CAB International, 1997), 28.

177 The book is *Affective Neuroscience: The Foundations of Human and Animal Emotions* (New York: Oxford University Press, 1998).

178 Darwin, *The Descent of Man and Selection in Relation to Sex*, 2nd ed. (1901), 192.

179 F. Darwin and A. C. Seward, *More Letters of Charles Darwin* (London: John Murray, 1903, vol. 1, p. 114). Note: "In a paper of penciled notes pinned into Darwin's copy of *Vestiges* occur the words 'Never use the word [sic] higher and lower.'"

180 Quoted in Jon Wynne-Tyson, *The Extended Circle: An Anthology of Humane Thought*, 2nd ed. (London: Sphere Books, 1990), 191.

182 See Bernard J. Baars, *The Cognitive Revolution in Psychology* (New York: The Guilford Press, 1986).

183 Pepperberg, *The Alex Studies*, 3.

184 N. Tinbergen, *The Study of Instinct* (Oxford: Oxford University Press, 1951), 4.

185 Antonio Damasio, *The Feeling of What Happens: Body, Emotion and the Making of Consciousness* (London: William Heinemann, 2000), 308.

186 See Mary Midgley's fine article on this topic, "Brutality and

Sentimentality," *Philosophy* 54 (1979): 385–9, where she points out that "notions like fear, anger, pleasure, etc., were not invented in or for an exclusively human world, any more than they were invented privately for inner pointing. They grew up in a thoroughly public world inhabited by many species, some of them constant companions of men. Species solipsism is no more convincing than the personal kind."

187 Kenneth Clark, *Animals and Men: Their Relationship as Reflected in Western Art from Prehistory to the Present Day* (London: Thames & Hudson, 1977), 45.

188 Zoophilia, as used by critics is loving animals too much. But originally the word simply meant being an opponent of cruelty to animals. Is that such a bad thing? Not if it counters zoophobia, a morbid fear of animals, or worse, anthropolatry, which the *Oxford English Dictionary* defines as worshiping man. If "sentimentalists" rely on emotions too much, they can answer that the critic is suffering from dogmatism, in the old-fashioned sense of using reason exclusively, and narrow-mindedness, also in the old-fashioned sense of lacking in wide sympathy. I do not wish to be antiscientific, but nor do I wish to fall prey to scientism. I don't want to be credulous and naive, believing everything is possible when it comes to love and animals, but nor do I wish to veer to the other extreme of being a radical skeptic and claiming that we can know nothing for certain.

189 This is the theme of a conference at King's College, London: "Putting Animal Sentience on the Educational Agenda," May 2003, in which the question of farm animal emotions was specifically addressed.

190 Margaret Floy Washburn, *The Animal Mind: A Text-book of Comparative Psychology*, The Animal Behavior Series, vol. 2 (New York: Macmillan, 1908), 7.

191 The April 28, 2003 Zogby poll found that seven out of ten likely voters in California favored passage of AB 732, a bill to protect animals from inhumane confinement on industrialized farms. This legislation required that pigs and calves be provided at least enough space to turn around, and it would prohibit the use of "veal crates" and "gestation crates." AB 732 passed the Public Safety committee on April 8th,

and was slated for a vote in the Agriculture Committee on Thursday, May 1. AB 732 was part of a growing nationwide effort aimed at prohibiting inhumane factory farming devices. This information comes from the Web site of Farm Sanctuary (www.farmsanctuary.org).

192 Peter Singer, "Animal Liberation at 30," *New York Review of Books*, May 15, 2003.

193 As for honey, I am somewhat ambivalent. In New Zealand, no bees are killed over the winter (as they are by large honey manufacturers in the United States, since it is cheaper to replace them than to feed them). Still, taking honey feels like theft to me. It is hard to give up what you are accustomed to, and my mother always believed that honey was a miracle food, so I was given lots of it as I was growing up. I am sure I will eventually relinquish it, just not today.

194 There are so many good books about both the history of vegetarianism and actually becoming one that it is difficult to choose. Let me suggest a few: A good overview, especially of the philosophical arguments, is Michael Allen Fox, *Deep Vegetarianism* (Philadelphia: Temple University Press, 1999). A good history by the food critic for the *Guardian* is Colin Spencer's *The Heretic's Feast: A History of Vegetarianism* (Hanover: University Press of New England, 1995). See also Paul R. Amato and Sonia A. Partridge, *The New Vegetarians: Promoting Health and Protecting Life* (New York: Plenum Press, 1989) and Peter Cox, *Why You Don't Need Meat* (London: Bloomsbury Publishing, 1992). See too John Lawrence Hill, *The Case for Vegetarianism: Philosophy for a Small Planet* (Lanham, Md.: Rowman & Littlefield, 1996). Still useful is the older *Vegetarianism: A Way of Life* by Dudley Giehl, with a foreword by Isaac Bashevis Singer (New York: Barnes & Noble Books, 1979). On veganism, see Kath Clements, *Why Vegan: The Ethics of Eating and the Need for Change* (London: Heretic Books, 1995).

195 See James Bovard, *ADM: A Case Study in Corporate Welfare*, Sept. 26, 1995. The full text is available online at www.cato.org.

196 A.V. Krebs operates the Corporate Agribusiness Research Project, P.O. Box 2201, Everett, WA 98203; e-mail avkrebs@earthlink.net; Web site: www.electricarrow.com/CARP

197 The information for this section comes from an excellent article by William Greider in *The Nation*, "The Last Farm Crisis." It is the feature story in the November 20, 2000 issue. I have also profited from reading the Web sites of the Organization for Competitive Markets (www.competitivemarkets.com) and the Organic Consumers Association (www.organicconsumers.org).

Bibliography

Items with a * were particularly helpful.

FARM ANIMALS IN GENERAL

Campbell, John R., and John F. Lasley, *The Science of Animals that Serve Mankind*, 2d ed. (New York: McGraw-Hill, 1975).

Delannoy, Dominique, *Animaux de la Ferme* (Losange: Artemis, 2000).

* Damerow, Gail, ed., *Barnyard in Your Backyard: A Beginner's Guide to Raising Chickens, Ducks, Geese, Rabbits, Goats, Sheep, and Cattle* (North Adams, Mass.: Storey Books, 2002).

Fox, Michael W., *Farm Animals: Husbandry, Behavior, and Veterinary Practice* (Baltimore: University Park Press, 1984).

Fraser, Allan, *Animal Husbandry Heresies* (London: Crosby Lockwood & Son, 1960).

Fraser, Allan, and D. M. Broom: *Farm Animal Behaviour and Welfare* (Oxford: CAB International, 1997).

Hafez, E. S. E., ed., *The Behaviour of Domestic Animals*, 3d ed. (London: Bailliére Tindall, 1975).

———, ed., *Reproduction in Farm Animals* (Philadelphia: Lea & Febiger, 1980).

Haynes, N. Bruce, *Keeping Livestock Healthy: A Veterinary Guide to Horses, Cattle, Pigs, Goats & Sheep*, 3d ed. (Pownal, Vt.: Storey Communications, 1985).

Isaac, Peter, *The Farmyard Companion* (London: Jill Norman & Hobhouse, 1981).

Spaulding, C. E., and Jackie Clay: *Veterinary Guide for Animal Owners* (Emmaus, PA: Rodale Press, 1998).

DOMESTICATION

Budiansky, Stephen, *The Covenant of the Wild: Why Animals Choose Domestication* (New York: William Morrow, 1992).

Caras, Roger A., *A Perfect Harmony: The Intertwining Lives of Animals and Humans Throughout History* (New York: Simon & Schuster, 1996).

* Clutton-Brock, Juliet: *A Natural History of Domesticated Mammals*, 2d ed. (Cambridge: Cambridge University Press [The Natural History Museum], 1999).

———, ed., *The Walking Larder: Patterns of Domestication, Pastoralism, and Predation* (London: Unwin Hyman, 1989).

Hemmer, Helmut, *Domestication: The Decline of Environmental Appreciation* (Cambridge: Cambridge University Press, 1990).

Houpt, Katherine Albro, *Domestic Animal Behavior for Veterinarians and Animal Scientists*, 2d ed. (Ames, Iowa: Iowa University State Press, 1992).

Hyams, Edward: *Working for Man: The Domestication of Animals* (Harmondsworth Middlesex: Penguin Books, 1975).

Newby, Jonica, *The Animal Attraction: Humans and Their Animal Companions* (Sydney: Australian Broadcasting Company, 1999).

Sauer, Carl O., *Agricultural Origins and Dispersal: The Domestication of Animals and Foodstuffs* (Cambridge, Mass.: M.I.T. Press, 1952).

Simmons, I. G., *Changing the Face of the Earth: Culture, Environment, History* (Oxford: Blackwell, 1989).

Stanford, Craig B., *The Hunting Apes: Meat Eating and the Origins of Human Behavior* (Princeton: Princeton University Press, 1999).

Urquhart, Judy, *Animals on the Farm: Their History from the Earliest Times to the Present Day* (London: Macdonald & Co., 1983).

* Zeuner, F. E., *A History of Domesticated Animals* (London: Hutchinson, 1963).

PIGS

Coffman, Steven, *How to Walk a Pig* (New York: Lyons & Burford, 1995).

Hedgepeth, William, *The Hog Book* (New York: Doubleday, 1978).

Holden, Philip, *Wild Pig in Australia* (Kenthurst, Australia: Kangaroo Press, 1994).

Hopf, Alice L., *Pigs Wild and Tame* (New York.: Holiday House, 1979).

Horwitz, Richard P., *Hog Ties: Pigs, Manure, and Mortality in American Culture* (New York: St. Martin's Press, 1998).

Hulme, Susan, *Book of the Pig* (Surrey: Spur Publications, 1979).

Malcolmson, Robert, and Stephanos Mastoris: *The English Pig: A History* (London: Hambledon Press, 1998).

* Meynhardt, Heinz: *Schwarzwild-Report: Mein Leben unter Wildschweinen* (Leipzig: Neumann Verlag, 1986).

Nissenson, Marilyn and Susan Jonas, *The Ubiquitous Pig* (New York: Harry N. Abrams, 1992).

Pond, Wilson G., and Katherine A. Houpt, *The Biology of the Pig* (Ithaca, N.Y.: Cornell University Press, 1978).

Porter, Valerie, *Pigs: A Handbook to the Breeds of the World* (Sussex: Helm Information, 1993).

Pukite, John, *A Field Guide to Pigs* (Helena, Montana: Falcon, 1999).

Rath, Sara, *The Complete Pig* (Stillwater, Minn.: Voyageur Press, 2000).

Schwenke, Karl, *In a Pig's Eye* (Chelsea, Vt: Chelsea Green Publishing Company, 1985).

* Sillar, F. C. and R. M. Meyler, *The Symbolic Pig: An Anthology of Pigs in Literature and Art* (Edinburgh: Oliver & Boyd, 1961).

Sonder, Ben, *Pigs & Wild Boars: A Portrait of the Animal World* (New York: Todtri Publications, 1998).

CHICKENS

* Davis, Karen: *More than a Meal: The Turkey in History, Myth, Ritual, and Reality* (New York: Lantern Books, 2001).

* ———, *Prisoned Chickens Poisoned Eggs: An Inside Look at the Modern*

Poultry Industry (Summertown, Tenn.: Book Publishing Company, 1996).

Druce, Clare, *Chicken & Egg: Who Pays the Price?*, with an introduction by Richard Adams (London: Merlin Press, 1989).

Grimes, William, *My Fine Feathered Friend* (New York: North Point Press, 2002).

Kruijt, J. P., *Ontogeny of Social Behaviour in Burmese Red Junglefowl (Gallus gallus spadiceus)* (Leiden: E. J. Brill, 1964).

* Lind, L. R. trans., *Aldrovandi on Chickens: the Ornithology of Ulisse Aldrovandi (1600)* (Norman, Okla.: University of Oklahoma Press, 1963).

Percy, Pam, *The Complete Chicken* (Stillwater, Minn.: Voyageur Press, 2002).

Porter, Valerie, *Domestic and Ornamental Fowl* (London: Penguin [Pelham Books], 1989).

Rogers, Lesley J., *The Development of Brain and Behaviour in the Chicken* (Oxford: CAB International, 1995).

* Smith, Page and Charles Daniel, *The Chicken Book* (San Francisco: North Point Press, 1982).

Staples, Tamara, Ira Glass: *The Fairest Fowl: Portraits of Championship Chickens* (San Francisco: Chronicle Books, 2001).

Wood-Gush, D. G. M., *The Behaviour of the Domestic Fowl* (London: Heinemann, 1971).

Cows

Carlson, Laurie Winn, *Cattle: An Informal Social History* (Chicago: Ivan R. Dee, 2001).

Geist, V., and F. Walther, eds., *The Behaviour of Ungulates and its Relation to Management*, 2 vols. (Morges, Switzerland: Conservation of Nature and Natural Resources, 1974).

Kiley-Worthington, Marthe, Susan de la Plain, *The Behaviour of Beef Suckler Cattle (Bos taurus)* (Boston: Birkhauser Verlag, 1983).

Lodrick, Deryck O., *Sacred Cows, Sacred Places: Origins and Survival of*

Animal Homes in India (Berkeley: University of California Press, 1981).

Lovenheim, Peter, *Portrait of a Burger as a Young Calf: The Story of One Man, Two Cows, and the Feeding of a Nation* (New York, Harmony Books, 2002).

Mourant, A. E., and F. E. Zeuner, eds., *Man and Cattle* (London: Royal Anthropological Institute of Great Britain & Ireland, 1963).

Porter, Valerie, *Caring for Cows* (London: Whittet Books, 1991).

———, *Cattle: A Handbook to the Breeds of the World* (London: Christopher Helm, 1991).

Rath, Sara, *About Cows* (Minocqua Wis.: NorthWord Press, 1987).

———, *The Complete Cow* (New York: Barnes & Noble, 1998).

Simoons, Frederick J., & Elizabeth S. Simoons, *A Ceremonial Ox of India: The Mithan in Nature, Culture and History: With Notes on the Domestication of Common Cattle* (Madison, Wis.: University of Wisconsin Press, 1968).

Webster, John, *Understanding the Dairy Cow* (Oxford: Blackwell Science, 1993).

SHEEP & GOATS

Auboiron, Bruno, and Gilles Lansard, *La transhumance et le berger: une tradition vivante* (Aix-en-Provence: Edisud, 1998).

Arthursson, Elizabeth: *Ewes & I* (London: Souvenir Press, 1988).

Chadwick, Douglas H., *A Beast the Color of Winter: The Mountain Goat Observed* (San Francisco: Sierra Club Books, 1983).

Clark, Bill, *High Hills and Wild Goats: Life Among the Animals of the Hai-Bar Wildlife Refuge* (Boston: Little, Brown & Co., 1990).

Ellison, Joan Jarvis, *Shepherdess: Notes from the Field* (West Lafayette, Ind.: Purdue University Press, 1995).

French, M. H., *Observations on the Goat* (Rome: Food and Agriculture Organization of the United Nations, 1970).

* Geist, Valerius, *Mountain Sheep: A Study in Behavior and Evolution* (Chicago: University of Chicago Press, 1971).

————, *Mountain Sheep and Man in the Northern Wilds* (Ithaca, N.Y.: Cornell University Press, 1975).

————, *Wild Sheep Country* (Minocqua, Wis.: NorthWord Press, 1993).

Hudson, W. H., *The Illustrated Shepherd's Life* (1910; reprint, London: The Bodley Head, 1987).

Ilefeldt, W. G., *Thoughts While Tending Sheep* (New York: Crown, 1988).

Jones, David Keith, *Shepherds of the Desert* (London: Elm Tree Books, 1984).

Jenkins, Marie M., *Goats, Sheep, and How They Live* (New York: Holiday House, 1978).

Mackenzie, David, *Goat Husbandry* (London: Faber & Faber, 1957).

Monson, Gale, and Lowell Sumner, eds., *The Desert Bighorn: Its Life History, Ecology, and Management* (Tucson, Ariz.: University of Arizona Press, 1981).

Müller, Hans Alfred, *Sheep* (New York: Barron, 1989).

Porter, Valerie, *Goats of the World* (Ipswich, U.K.: Farming Press, 1996).

Reynes, Nicole, and Christophe Latour, *Moutons & bergers* (Paris: Editions Rustica, 2000).

Schaller, George B., *Stones of Silence: Journeys in the Himalaya* (New York: Viking Press, 1979).

Seymour, John, *The Shepherd* (London: Sidgwick & Jackson, 1983).

Thomas, J. F. H., *Sheep* (London: Faber & Faber, 1954).

* Townend, Christine, *Pulling the Wool: A New Look at the Australian Wool Industry* (Sydney: Hale & Ironmonger, 1985).

DUCKS & GEESE

* Armstrong, Edward A., *Bird Display and Behaviour: An Introduction to the Study of Bird Psychology*, rev. ed. (1942; reprint, New York: Dover, 1965).

————, *The Life & Lore of the Bird in Nature, Art, Myth, and Literature* (New York: Crown Publishers, 1975).

* Bent, Arthur Cleveland, *Life Histories of North American Wild Fowl:*

Duck, Geese, Teals, Mergansers, Eiders, Swans, Scoters and Others, 2 vols. (1925; reprint, New York: Dover, 1987).

Brooke, Michael, and Tim Birkhead, *The Cambridge Encyclopedia of Ornithology* (New York: Cambridge University Press, 1991).

Collias, N. E., and E. C. Collias, *Nest Building and Bird Behavior* (Princeton: University Press, 1984).

Delacour, Jean, *The Waterfowl of the World*, illustrated by Peter Scott., vol. 1 (London: Country Life Ltd., 1954).

Fiennes, William, *The Snow Geese* (London: Picador, 2002).

Gill, Frank B., *Ornithology* (New York: W. H. Freeman & Co., 1990).

Gilliard, E. Thomas, *Living Birds of the World* (New York: Doubleday & Co., 1958).

Giorgetti, Anna, *Ducks: Art, Legend, History* (Boston: Little, Brown & Co., 1991).

Hochbaum, H. Albert, *Travels and Traditions of Waterfowl* (Minneapolis: University of Minnesota Press, 1955).

Johnsgard, Paul A., *Handbook of Waterfowl Behaviour* (London: Constable & Co., 1965).

Kastner, J., *A World of Watchers: An Informal History of the American Passion for Birds* (San Francisco: Sierra Club, 1986).

Kear, Janet, *Man and Wildfowl* (London: T. & A. D. Poyser, 1991).

Lack, David, *Evolution Illustrated by Waterfowl* (Oxford: Blackwell Scientific Publications, 1974).

Lishman, William, *Father Goose: The Adventures of a Wildlife Hero* (Toronto: Little, Brown & Co.: 1995).

Lorenz, Konrad, *Here Am I: Where Are You: The Behavior of the Greylag Goose* (New York: Harcourt Brace Jovanovich, 1991).

———, *The Year of the Greylag Goose* (New York: Harcourt Brace Jovanovich, 1978).

Luttmann, Rick, and Gail Luttmann, *Ducks & Geese in Your Backyard* (Emmaus, Pa.: Rodale Press, 1978).

Soothill, Eric and Peter Whitehead, *Wildfowl of the World* (Poole, U.K.: Blandford Press, 1978).

Tangye, Derek, *A Drake at the Door* (London: Michael Joseph, 1963).

Tomlinson, David, *Ducks* (London: Whittet Books, 1996).

* Welty, Joel Carl, *The Life of Birds* (New York: Alfred A. Knopf, 1968).

Wilson, B., ed., *Birds: Readings from Scientific American* (San Francisco: W. H. Freeman, 1980).

FACTORY FARMING

Bauston, Gene, *Battered Birds, Crated Herds: How We Treat the Animals We Eat* (Watkins Glen, N.Y.: Farm Sanctuary: 1996).

Coats, C. David, *Old MacDonald's Factory Farm: The Myth of the Traditional Farm and the Shocking Truth about Animal Suffering in Today's Agribusiness* (New York: Continuum, 1989).

Comstock, Gary, ed., *Is There a Moral Obligation to Save the Family Farm?* (Ames, Iowa: Iowa State University Press, 1987).

Harrison, Ruth, *Animal Machines* (London: Vincent Stuart, Ltd., 1964).

* Mason, Jim, and Peter Singer, *Animal Factories*, rev. ed (New York: Harmony Books, 1990).

McKenna, Carol, *Forced Feeding: An Inquiry into the Welfare of Ducks and Geese Kept for the Production of Foie Gras* (London: World Society for the Protection of Animals, 2000).

Nestle, Marion, *Food Politics: How the Food Industry Influences Nutrition and Health* (Berkeley: University of California Press, 2002).

Rifkin, Jeremy, *Beyond Beef: The Rise and Fall of the Cattle Culture* (New York: Dutton, 1992).

Rollin, Bernard E., *Farm Animal Welfare: Social, Bioethical, and Research Issues* (Ames, Iowa: Iowa State University Press, 1995).

Schell, Orville, *Modern Meat: Antibiotics, Hormones, and the Pharmaceutical Farm* (New York: Random House, 1984).

* Singer, Peter, *Animal Liberation*, 2d ed. (New York: New York Review of Books, 1990).

Stevenson, Peter, *A Far Cry from Noah: The Live Export Trade in Calves, Sheep and Pigs* (London: Green Print, 1994).

Strange, Marty, *Family Farming: A New Economic Vision* (San Francisco: Institute for Food and Development Policy, 1988).

VEGETARIANISM

Amato, Paul R., and Sonia A. Partridge, *The New Vegetarians* (New York: Plenum Press, 1989).

Belasco, Warren J., *Appetite for Change: How the Counterculture Took on the Food Industry 1966–1988* (New York.: Pantheon Books, 1989).

* DeGrazia, David, *Taking Animals Seriously: Mental Life and Moral Status* (New York: Cambridge University Press, 1996).

* Fiddes, Nick, *Meat: A Natural Symbol* (London and New York: Routledge, 1991).

Fox, Michael Allen, *Deep Vegetarianism* (Philadelphia: Temple University Press, 1999).

* Gellatley, Juliet with Tony Wardle, *The Silent Ark: A Chilling Expose of Meat: The Global Killer* (London: HarperCollins, 1996).

Giehl, Dudley, *Vegetarianism: A Way of Life.*, with a forword by Isaac Bashevis Singer (New York: Barnes & Noble, 1979).

Hausman, Patricia: *Jack Sprat's Legacy: The Science & Politics of Fat & Cholesterol* (New York: Center for Science in the Public Interest, Richard Marek Publishers, 1981).

Hulse, Virgil, *Mad Cows and Milk Gate* (Phoenix, Oreg.: Marble Mountain Publishing, 1996).

Hill, John Lawrence, *The Case for Vegetarianism: Philosophy for a Small Planet* (Lanham, Md.: Rowman & Littlefield Publishers, 1996).

Lappé, Frances Moore, *Diet for a Small Planet*, 2d rev. ed. (New York: Ballantine Books, 1987).

Lyman, Howard F., with Glen Merzer, *Mad Cowboy: Plain Truth from the Cattle Rancher Who Won't Eat Meat* (New York: Scribner, 1998).

Marcus, Erik, *Vegan: The New Ethics of Eating* (Ithaca, N.Y.: McBooks Press, 1998).

* Pluhar, Evelyn B., *Beyond Prejudice: The Moral Significance of Human and Nonhuman Animals* (Durham, N.C.: Duke University Press, 1995).

Robbins, John, *Diet for a New America* (Walpole, N.H.: Stillpoint Publishing, 1987).

* Sorabji, Richard, *Animal Minds & Human Morals: the Origins of the Western Debate* (Ithaca, N.Y.: Cornell University Press, 1993).

Spencer, Colin, *The Heretic's Feast: A History of Vegetarianism* (Hanover: N.H.: University Press of New England, 1995).

Sussman, Vic, *The Vegetarian Alternative: A Guide to a Healthful and Humane Diet* (Emmaus, Pa.: Rodale Press, 1978).

Tansey, Geoff and Joyce D'Silva, eds., *The Meat Business: Devouring a Hungry Planet* (London: Earthscan Publications, 1999).

FOOD ISSUES

Critser, Greg, *Fat Land: How Americans Became the Fattest People in the World* (Boston: Houghton Mifflin Co., 2003).

Fernández-Armesto, Felipe, *Near a Thousand Tables: A History of Food* (New York: The Free Press, 2002).

Rhodes, Richard, *Farm: A Year in the Life of an American Farmer* (Lincoln, Nebr.: University of Nebraska Press, 1989).

* Schlosser, Eric, *Fast Food Nation: The Dark Side of the All-American Meal* (Boston: Houghton Mifflin Co., 2001).

SOME OF MY FAVORITE ANIMAL BOOKS
FOR BACKGROUND READING

Bekoff, Marc, ed., *Smile of a Dolphin: Remarkable Accounts of Animal Emotions*, with a foreword by Stephen Jay Gould (New York: Discovery Books, 2000).

Cartmill, Matt, *A View to a Death in the Morning: Hunting and Nature Through History* (Cambridge: Harvard University Press, 1993).

Griffin, Donald R., *Animal Minds: Beyond Cognition to Consciousness*, rev. ed. (Chicago: University of Chicago Press, 2001).

Heinrich, Bernd, *Mind of the Raven: Investigations and Adventures with Wolf-Birds* (New York: Cliff St. Books, 1999).

Hutto, Joe, *Illumination in the Flatwoods: A Season with the Wild Turkeys* (New York: Lyons & Burford, 1995).

Mason, Jim, *An Unnatural Order: Uncovering the Roots of our Domination of Nature and Each Other* (New York: Simon & Schuster, 1993).

Midgley, Mary, *Beast and Man: the Roots of Human Nature* (Ithaca, N.Y.: Cornell University Press, 1978).

Regan, Tom, *The Case for Animal Rights* (Berkeley: University of California Press, 1983).

Rollin, Bernard E., *The Unheeded Cry: Animal Consciousness, Animal Pain and Science* (Oxford; New York: Oxford University Press, 1989).

Ryden, Hope, *Lily Pond: Four Years with a Family of Beavers* (New York: William Morrow & Co., 1989).

Scully, Matthew, *Dominion: The Power of Man, the Suffering of Animals, and the Call to Mercy* (New York: St. Martin's Press, 2002).

Thomas, Keith: *Man and the Natural World: A History of the Modern Sensibility* (New York: Pantheon Books, 1983).

Turner, E. S., *All Heaven in a Rage* (London: Michael Joseph, Ltd., 1964).

Wise, Steven M.: *Rattling the Cage: Towards Legal Rights for Animals* (New York: Perseus Books, 1999).

Resources

First go to: www.sanctuaries.org, for a list of the best and most trust-worthy animal sanctuaries. It is constantly updated.

PETA (People for the Ethical Treatment of Animals): www.peta.org

United Poultry Concerns: www.upc-online.org

Mercy for Animals: www.mercyforanimals.org

Compassion in World Farming: www.ciwf.co.uk

Compassion over Killing: www.cok.net

Vegan Action: www.vegan.org

Humane Society of the United Sates: www.hsus.org

Viva! USA: www.vivausa.org. (In the U.K. it is: www.viva.org.uk)

SAFE (Save Animals from Exploitation): www.safe.org.nz

In Defense of Animals: www.idausa.org

ANIMAL SANCTUARIES

Animal Place, near Davis, California www.animalplace.org

Farm Sanctuary, in Orland, California and Watkins Glen, New York.: www.farmsanctuary.org

Wilderness Ranch, in Loveland, Colorado: www.wildernessranch.org

Poplar Springs Animal Sanctuary, in Poolesville, Maryland: www.-animalsanctuary.org

Black Beauty Ranch, in Murchison, Texas: www.blackbeautyranch.org

A Film About
This Book

Award-winning filmmaker Stanley Minasian has made a film about the making of this book, called *The Emotional World of Farm Animals with Jeffrey Moussaieff Masson*. A film by Earth Views Productions and Animal Place, 2003. Directed by Stanley M. Minasian. Executive Producer: Kim Sturla. For copies contact: Animal Place, 3448 Laguna Creek Trail, Vacaville, CA 96588. Tel. (707) 449-4814, fax 449-8775. See: www.animalplace.org. The film was funded by Glaser Progress Foundation and Leonard Bosack and Bette Kruger Foundation.

Index